Associations and Other Groups in Science

Associations and Other Groups in Science: An Historical and Contemporary Perspective

Edited by

Ana Delicado

CAMBRIDGE
SCHOLARS
PUBLISHING

Associations and Other Groups in Science:
An Historical and Contemporary Perspective,
Edited by Ana Delicado

This book first published 2013

Cambridge Scholars Publishing

12 Back Chapman Street, Newcastle upon Tyne, NE6 2XX, UK

British Library Cataloguing in Publication Data
A catalogue record for this book is available from the British Library

ISBN (10): 1-4438-5224-4, ISBN (13): 978-1-4438-5224-1

TABLE OF CONTENTS

LIST OF TABLES

INTRODUCTION

ANA DELICADO

Scientific societies or associations are quite an under-researched issue. Science studies have historically paid much more attention to non-formalised collectives in science: the "republic of science" of Polanyi (1962), the "scientific community" of Hagstrom (1965) and Merton (1973), the "invisible colleges" of Crane (1972), the "scientific field" of Bourdieu (1975), or the "transepistemic arenas" of Knorr-Cetina (1982). Theories of the Mode 2 production of knowledge (Gibbons et al. 1995; Nowotny, Scott and Gibbons 2001) postulate that in contemporary science there is a wider array of institutions taking part in the construction of scientific knowledge, but scientific associations are barely mentioned. Recent literature has devoted considerable attention to the "triple helix" formed by universities, government and industry (Leydesdorff and Etzkowitz 1996; Shinn 2002), but leaving out all other actors involved in science.

Most of the existing literature on scientific associations focuses mainly on the historical dimension of scientific societies, such as their role in the birth of modern science (see Merton 1938 or Shapin 1996, for instance) or the genealogy of individual institutions, such as the Royal Society (Hunter 1989). Just a few noteworthy exceptions can be found: for instance, an in-depth case study of the Society of German Chemists authored by Rilling (1986), Schimank's (1988) survey of German scientific associations, the studies on the development of international scientific associations by Shofer (1999, 2003). Some published research also deals with particular aspects of scientific associations' activities, such as publishing (Levitan 1979), policy advice (Teich 2002), ethical regulation (Levine and Iutcovich 2003), award giving (Crosland and Galvez 1989), public understanding of science (Rogers 1981), or membership profiles (Mackie 2000).

And yet, scientific societies in Europe seem to be gaining momentum both in individual countries (see, for instance, Guinovart 2009 on the Spanish Federation of Scientific Societies) and on a transnational level. This is apparent, for instance, in the Initiative for Science in Europe - ISE, a platform of European learned societies and scientific organisations created in 2004, which led to the creation of the European Research Council (Zaragoza 2007).

Overall, there seemed to be a gap in the social studies of science concerning the role of scientific societies in contemporary science. In 2008, a team of social scientists sought to bridge this gap with a research project,[1] focusing on the case of the Portuguese scientific system, which intended to explore both the functions and activities of scientific societies and the associative behaviour and representations of scientists.

This research project developed a two-tiered approach. First, a census of existing Portuguese scientific societies was conducted, followed by an online survey. A sample of organisations was then selected for an in-depth analysis, based on document collection, interviews with members of the governing boards, observation at events and a survey of their members. The purpose of these procedures was to gather information on the activities of the associations, their internal functioning, their resources and handicaps, how they are articulated with other associations and organisations in the scientific system and what roles they play in some specific areas: policy advice, science and business links, science and society relations, the social responsibility of scientists.

Second, an internally diverse sample of scientists working in Portuguese institutions was surveyed, in order to inquire on their associative behaviour and representations, at both national and international levels. Here, the main purpose was to ascertain the role associations play in scientific careers, in research practices and in mediating the relation between individual scientists and wider collective structures (their discipline, the scientific community, the government, industry).

This research project relied on a wide definition of scientific societies (or associations), which encompassed private non-profit institutions devoted to professional and disciplinary aims, labour activism, public engagement with science, links with innovation and industry, diaspora networks. The particular topic of scientific associations was expected to stimulate the reflection and discussion of wider issues in science, such as the need for building bridges within the scientific community and with society at large. Although it was focused on a national case, it is deemed relevant to the international community of social studies of science, since

---

[1] *SOCSCI Scientific Societies in Contemporary Science,* funded by the Portuguese Foundation for Science and Technology (PTDC/CS-ECS/101592/2008), hosted by the Institute of Social Sciences of the University of Lisbon, in collaboration with CIES-IUL and SOCIUS-ISEG.

it has yielded results that are valid across borders and can be used to integrate comparative research with other scientific systems. And though scientific societies may occupy a relatively small place in a field dominated by other institutions, this research meant to contribute to highlight the work of these organisations, bringing science and society a little closer, and to inform decision-making at several levels (science and technology policies, the strategies and actions of associations and individual scientists).

In June 2012, in the final stage of this research project, a conference was held to showcase its main results but also to give an opportunity for other researchers to present their work in this and other connected areas.[2] A call for papers was issued and proposals from historians, anthropologists and sociologists, working in many different areas, poured in. In what can be a sign of the lack of research on this issue, despite the fact that the call was disseminated internationally (in English, in several international mailing lists and newsletters) barely any proposals from other than Portuguese scholars were received. This book is the result of that conference. It brings together some of the papers presented on that day, revised and expanded in order to accommodate a fruitful dialogue between them.

The book is divided in two parts, reflecting a temporal and thematic chasm. The first part concerns historical studies focused mostly on the part that associations (and other groups) played in the development of particular scientific disciplines in the nineteenth and twentieth century; the second part is dedicated to contemporary analyses that chiefly highlight the engagement between associations and wider publics.

The book opens with a chapter by Cristiana Bastos on the most traditional of scientific associations: the Royal Academy of Sciences of Lisbon, a congener of similar institutions throughout Europe in the eighteenth century. Rather than opting for a conventional account of the origins of the Academy (which already abound), Bastos develops a highly innovative approach, following a thread that leads us from objects to people, from the South Pacific to Europe, from politics to literature and gender. Focusing on two characters that played a major role in the early years of the Academy, the Abbot Correia da Serra and the Countess Leonor de Almeida, who met only briefly on a coach trip between Vienna

[2] The conference took place at the National Museum of Natural History and Science of the University of Lisbon, a landmark of Portuguese science and the birthplace of several of the most prominent scientific societies in the country. We are grateful to Marta Lourenço for welcoming us in the Museum.

and Lisbon, this chapter offers an engaging interpretation of scientific networks and the circulation of ideas during the Enlightenment.

Chapter Two, on the Association of Portuguese Archaeologists, by Ana Cristina Martins, also deals with material culture in museums but in this case in connection with the institutionalisation of archaeological sciences. Against the backdrop of the development of archaeology in Europe and in Portugal, this chapter highlights the role the Association and its mentor, Possidónio da Silva, played in the field of heritage preservation throughout the nineteenth century and in the first decades of the twentieth century. It traces the changes the Association underwent, as well as the impact it had on national and local government policy.

Hugo Pereira's chapter concerning the Civil Engineers Association delves as well on the Association's influence over government policy, this time on the particular issue of railway development. The chapter charts the growth of railroads in Portugal in the later part of the nineteenth century, examining the lobbying strategies of the Association, the technical advice it proffered through its journal and the part it played in generating a debate on planning issues. The chapter also emphasises the conflicting interests within the Association and the predominance of centralised, political decision-making over expertise-based judgments.

Chapter Four marks a return to a classical nineteenth century type of scientific association: the Geographical Society of Lisbon. Rui Pinto touches upon some of the key moments of the early days of the Society, strongly influenced by the colonial power play in Europe: expeditions in Africa, exhibitions of exotic products, training of colonial officers, political statements at critical moments.

The following chapter, by Ricardo Moreira, moves the analysis of scientific associations forward into the twentieth century. Focusing on the particular case of the disciplines of oncology and radiology, this chapter connects the origins of several scientific societies to teams and individual actors working on cancer research and treatment. Modern practices in medical research, laboratory work or hospital organisation influenced the internal organisation of scientific societies. A new type of association, of a professional nature, also emerges in this context, reflecting the trend of medical specialisation and responding to the growing anxieties of medical practitioners concerning the risks and dangers of radiation.

Chapter Six, by Dulce Freire, concludes this part with an account of the origins and activities of a research organisation which sought to promote the modernisation of Portuguese agriculture, under the aegis of a private foundation and negotiating the pitfalls of a science-unfriendly authoritarian regime.

Part II of the book begins with three chapters resulting from the SOCSCI project. Chapter Seven, by Raquel Rego, claims that despite their outstanding growth in recent years, scientific associations still attain low levels of professionalisation and institutionalisation, which in turn have a bearing on their limited ability to exert political influence. It pays particular attention to associations of a professional nature, formed around new occupational groups in science.

The next chapter analyses the participation of researchers in these professional associations, aiming to characterise their membership base, motivations for belonging and modes of participation. Luis Junqueira draws from the literature on labour movements in universities and academic capitalism, as well as the growth of the Portuguese science and technology system, to explain the rise of these associations.

Cristina Palma Conceição and Ana Delicado's chapter examines the science communication activities of the other two types of association: scientific societies (mostly of a disciplinary nature) and scientific dissemination associations (astronomy clubs, nature groups, science promotion NGOs). In the former, public understanding of science is becoming an increasingly important issue, though part of a wider array of activities and functions, much as a result of the competition with international associations that depleted the traditional roles of these societies. The latter type tends to involve a more diverse range of members (scientists, teachers, students) and a wider variety of dissemination activities, much more oriented towards public engagement with science (and even the production of new scientific knowledge). Its growth can be an indicator of an increasing link between science and society, as well as of the growing professionalisation and specialisation in science communication.

The following two chapters offer two prime examples of this trend, by presenting the case of two associations devoted to scientific dissemination. Chapter Ten focuses on the role of a local association in promoting a fairly neglected field in academia, speleology. Based mostly on contributions from amateurs, AESDA reproduces most of the practices carried out in the scientific field: conducting systematic research, collaborating with other teams, presenting paper at conferences, submitting articles to peer-reviewed journals, publishing its own periodical, donating interesting specimens to museums.

The chapter of Sandra Assis and Vanessa Campanacho examines the creation of the Group of Studies in Human Evolution, a non-profit association constituted in the year 2005 by a group of students with the aim of disseminating scientific knowledge among the general public. The authors describe the comprehensive array of activities the Group has developed over the years, fulfilling its mission to bridge the sometimes wide gap between academic institutions and society at large.

The book concludes with an analysis of individual and collective "scientific activism" in nuclear controversies, by Ana Delicado. The chapter scrutinises the participation of scientists and associations in the recurring debate on resorting to nuclear energy in Portugal, connecting it to trends in scientific advice and governance and to the rise of the environmental movement. The events narrated in this chapter reveal the tensions within the scientific community, but also the alliances that are formed with external actors when matters of societal relevance are at stake.

Though the role of associations in science is the main thread that binds the chapters in this book, a few other common strands can be found. Several chapters show the role charismatic leadership can have in promoting and mobilising collective action: Correia da Serra in the Academy of Sciences (Chapter One), Possidónio da Silva in the Association of Portuguese Archaeologists (Chapter Two), Luciano Cordeiro in the Lisbon Geographical Society (Chapter Four), Mark Atthias in the Society of Natural Sciences (Chapter Five), Mário de Azevedo Gomes and Henrique de Barros at CEEA (Chapter Six). The importance of international networks is a recurring theme, demonstrating that Portuguese science and its associations have not evolved in isolation, but rather benefited from the circulation of knowledge afforded by forging links with foreign organisations and scholars. Another issue that emerges in several chapters is the sometimes fraught relations between the scientific community and government. Either asserting their labour rights or proffering technical advice, researchers clamour to be heard in political decision-making. However, politicians and officials seem to seldom welcome these contributions, upholding a centralised and hierarchical form of governance. Finally, some chapters also illustrate the importance of non-human actors, so dear to the Actor-Network Theory. In some of the stories told here, material objects take centre stage and sway the relations around them: museum artefacts, ruins, railways, hospital buildings, caves, bones.

Thus, scientific associations, though a fairly neglected issue in science studies, do offer the opportunity of examining broader matters concerning the functioning of the scientific system and its connections with the social context in which it operates.

Bibliography

Bourdieu, Pierre. 1975. "La Spécificité du Champ Scientifique et les Conditions Sociales du Progrès de la Raison." *Sociologie et Societés* 7 (June 2011): 91–118.

Crane, D. 1972. *Invisible colleges. Diffusion of knowledge in scientific communities.* Chicago: Chicago University Press.

Crosland, M. & A. Gálvez. 1989. "The Emergence of Research Grants Within the Prize System of the French Academy of Sciences, 1795-1914." *Social Studies of Science* 19: 71–100.

Gibbons, M. *et al.* 1994. *The New Production of Knowledge. The Dynamics of Science and Research in Contemporary Societies.* Londres: Sage.

Guinovart, J. J. 2009. "Mind the Gap: Bringing Scientists and Society Together." *Cell* 137 (5) (May): 793–5.

Hagstrom, W. O. 1965. *The Scientific Community.* New York: Basic Books.

Hunter. 1989. *Establishing the New Science: The Experience of the Early Royal Society.* Woodbridge: Boydell & Brewer Ltd.

Knorr-Cetina, K. 1982. "Scientific community or transepistemic arenas of research? A critique of quasi-economic models of science". *Social Studies of Science*, 12, 101-130.

Levine, F. J., & J. M. Iutcovich. 2003. "Challenges in studying the effects of scientific societies on research integrity". *Science and engineering ethics* 9 (2): 257-268.

Levitan, K. B. 1979. "Scientific societies and their journals: Biomedical scientists assess the relationship". *Social Studies of Science*, 9 (3): 393-400.

Leydesdorff, L. & H. Etzkowitz. 1996. "Emergence of a triple helix of University-Industry-Government relation". *Science and Public Policy*, 23: 279-286.

Mackie, R. 2000. "'But What Is a Chemical Engineer?': Profiling the Membership of the British Institution of Chemical Engineers, 1922–1956." *Minerva* 38: 171–199.

Merton, R. K. 1938. "Science, technology and society in seventeenth century England". *Osiris*, 4: 360-632.

—. 1973. *The sociology of science: theoretical and empirical investigations.* Chicago: Chicago University Press.

Nowotny, H., P. Scott & M. Gibbons. 2001. *Rethinking Science: Knowledge in an Age of Uncertainty.* Cambridge: Polity.

Polanyi, M. 1962. "The Republic of Science: Its Political and Economic Theory". *Minerva*, 1, 54-74.

Rilling, R. 1986. "The structure of the Gesellschaft Deutscher Chemiker (Society of German Chemists)". *Social Studies of Science*, 16 (2): 235-260.

Rogers, C. L. 1981. "Science information for the public: The role of scientific societies". *Science, Technology & Human Values*, 6 (36): 36-40.

Schimank, U. 1988. "Scientific associations in the German research system – Results of an empirical study". *Knowledge in Society*, 1 (2): 69-85.

Schofer, E. 1999. "Science Associations in the International Sphere 1875-1990: The Rationalization of Science and the Scientization of Society." In *Constructing World Culture: International Nongovernmental Organisations Since 1875*, ed. J. Boli & G. M. Thomas, Stanford CA: Stanford University Press. 249–266.

—. 2003. "International Science Associations, 1870-1990." In *Science in the Modern World Polity: Institutionalization and Globalization*, ed. G. S. Drori, J. W. Meyer, F. O. Ramirez & E. Schofer, Stanford CA: Stanford University Press, 81–99.

Shapin, S. 1996. *The scientific revolution*. Chicago: Chicago University Press.

Shinn, T. 2002. "The triple helix and new production of knowledge: prepackaged thinking on science and technology." *Social Studies of Science,* 32: 599-614.

Teich, A. 2002. "AAAS and public policy: speaking softly and carrying a medium-sized stick". *Technology in Society* 24 (1-2): 167-178.

Zaragoza, F. M. 2007. "Promotion of basic research in the European Union: the European Research Council (ERC) and the Initiative for Science in Europe (ISE)". *Anales de la Real Academina Nacional de Medicina,* 124 (3): 559-567.

PART I

THE ROLE OF ASSOCIATIONS IN THE DEVELOPMENT OF SCIENTIFIC DISCIPLINES

CHAPTER ONE

OF OBJECTS, NETWORKS, POLITICS AND POETRY: THE ROYAL ACADEMY OF SCIENCES OF LISBON IN THE CONTEXT OF THE ENLIGHTENMENT

CRISTIANA BASTOS

Objects of curiosity

Well-kept in the historical building of the Academy of Sciences in Lisbon, a number of artefacts from around the world co-exist with books, paintings, furniture and other rare objects from a time when knowledge, wealth, pomp and some degree of cosmopolitanism went together. Among them there is a small collection that includes a helmet, masks, hooks, oars and some other artistically crafted objects from the islands of Hawaii and the coast of Vancouver. These objects piqued the visitor's curiosity, not only due to their beauty and presumed value, but also, and foremost, for their unusual origin. Tinglit, Nootka and Kwakiutl are familiar names to anthropologists, but their artwork remains exclusive of very few international collections. They are absent from Portuguese museums. The same goes for Hawaiian peoples and their artefacts. None of them belonged to the far-reaching Portuguese networks in maritime trade, political relations or military conquest of colonial times, and from which the Portuguese brought back the goods, knowledge, crafts and artwork that compose existing ethnographic collections. Most of the ceremonial masks, religious artefacts, religious statuettes or domestic utensils that exist in Portuguese museums come from Africa or, to a lesser degree, from Asia,

South America or Melanesia/Timor. There are scarcely any objects from the northern regions of America, Asia or Europe.[1]

Maybe for that reason, the Northern Pacific objects of the Academy of Sciences had been initially classified as African, and for over a century remained misclassified, indistinguishable from many other museum objects, as if metonymically affected by the majority. Only towards the end of the twentieth century were they correctly identified and given appropriate museological care.[2] They were placed in special display cases in a front room at the Academy of Sciences, and can be visited on special occasions, but not photographed. The reasons and routes that brought them to Lisbon are still a matter of speculation. When did they arrive, how did they travel, through whose networks, and under what circumstances?

The travels of Captain Cook and his objects

My first acquaintance with the artefacts dates to a special visit to the *Aula Maynense* of the Academy of Sciences in March 2010, where I also became familiar with the oral history running in the institution: the objects had been collected by Captain Cook on his third and last trip around the world, made their way to England after his death, and from England were transferred to Portugal due to the connections of one of the Academy members.[3]

It is widely known that Captain Cook lost his life in Hawaii in 1779, during his third great voyage after exploring the coast of British Columbia and searching for the Behring passage. He first arrived in Hawaii during the Lono god festivals; whether he was taken for Lono himself, treated as a god, or something else, constitutes a matter of debate among

---

[1] For a wider discussion about Portuguese museums and collections, see Brigola (2003), Cantinho Pereira (2005) and Delicado (2009).

[2] A special thanks to Manuela Cantinho Pereira, PhD and specialist of colonial collections, for sharing her valuable knowledge with the author and research assistant Patrick Figueiredo on an interview at the Lisbon Geographical Society in Lisbon, 2011.

[3] The official page of the Museu Maynense at the Academy of Sciences states that among its ethnographic collections are “the precious, high quality and partly unpublished collection from British Columbia and Hawaii obtained in the third and last expedition of James Cook and transferred to the Academy thanks to Sir Joseph Banks (President of the Royal Society and a member of our Academy) and his friend the Abbott Correa da Serra” (http://www2.acad-ciencias.pt/joomla/index.php?option=com_content&view=article&id=255&Itemid=81, accessed on July 29, 2013).

anthropologists—who also use his case to engage in further discussions about issues as diverse as first encounters, human rationality, European culture and the art of butchery (Sahlins 1985, 1995; Obeyesekere 1997). The fact is that after having first left Hawaii in glory, his unexpected return to the island had dire consequences: he was killed and his body was ritually disassembled.

When the vessel *Endeavour* returned to England leaving Captain Cook's body parts behind, it is likely that there were quite a few artefacts on board, both from Hawaii and from British Columbia. It is also likely that they landed in England. Yet there is no evidence, so far, that the Lisbon Academy objects were among them, nor is it clear how they would have travelled to Lisbon. With some speculation and creative thinking, oral history suggests that the objects travelled from one place to the other due to the special friendship between two important men of science, one in London, the other in Lisbon.

One of those men was the famous naturalist Sir Joseph Banks (1743-1820). Banks had accompanied Captain Cook on his first voyage around the world on board of the *Endeavour* (1768-1771). Although the Cook-Banks connection prevails in historical and popular imagination, Banks did not participate in Captain Cook's second trip, aboard the *Resolution* (1772-1775), nor in the third, back on the *Endeavour* (1776-1779). He nonetheless remained at the centre of knowledge-making of his time. He surveyed, identified, collected and systematised knowledge about different species; he supervised their cultivation at Kew Gardens (Desmond 1995, Gascoigne 1998). He was at the core, and he had a vast network of collaborators, interlocutors and friends collecting data in numerous places.

The other man of science mentioned in the legend was the Portuguese naturalist Correia da Serra (1751-1823), one among Joseph Banks' friends and interlocutors, and one of the founders of the Royal Academy of Sciences in Lisbon in 1779 (Beale Davis 1993, Teague 1997, Carneiro *et al.* 2000, Diogo *et al.* 2001, Simões *et al.* 2006)

Objects and scientific networks

Rather than the less noble and more common reasons of colonial plunder or trade, friendship and intellectual camaraderie in the age of the Enlightenment explained the presence of these unusual pieces in Lisbon. They travelled to Lisbon as a gift, after a series of transactions involving several scientific celebrities; an attractive tale of high pedigree, in sum. Collected *in loco* by no less than Captain Cook and traveling after his death in Hawaii, the pieces would have made their way to Europe and were later given by Sir Joseph Banks, Captain Cook's associate and famous naturalist, to the Portuguese naturalist Correia da Serra, as part of the exchanges that sustained their friendship. If we apply Marcel Mauss' theories about "the gift" (Mauss 1925, 1990), we see that after this journey the objects possessed an overwhelming abundance of *Hau,* enriched by the vitality and notoriety of each of their owners, givers and receivers. Captain Cook, Joseph Banks and Correia da Serra had imbued their personas in the wonderful, rare and authentic masks, ceremonial helmet, hooks, oars and ornaments kept at the Academy.

The story seemed too good to be true: Banks at the centre of Enlightenment networks, the objects physically traveling from the islands of Hawaii and the shores of Vancouver to the city of Lisbon, via Banks, with Cook at one end and Serra on the other. Whether true or false, the story led me from objects to relationships and to a range of questions about the networks that at that time brought together naturalists and things, projects, research interests, sites on earth, flows of knowledge; about the insertion of the Lisbon Academy of Sciences in the wider Enlightenment networks; about the elusive Correia da Serra, who spent two-thirds of his life abroad and yet was one of the major contributors to science in Portugal; about his connections to Joseph Banks and other key figures of the Enlightenment, in France, England and the United States.

As much as the masks, oars, and helmet enchanted me, the questions around them enchanted me even more. I followed the threads that would lead me from the objects to the networks, trends, scientists and institutions of the Enlightenment. I also came upon unexpected scenarios involving not only science and politics, but also literature and poetry, class and gender , Freemasonry and religion, police and persecutions. The beautiful Kwakiutl objects and Cook's adventures in Hawaii were just the beginning of an exploration that led me away from the Pacific and into the heart of Europe's tensions at the end of the eighteenth century. While the route the objects took to Lisbon is still unclear and the subject of research by qualified scholars,[4] and while there is no support for the speculative hypothesis that they came to Lisbon via the Banks-Serra connection,[5] one

[4] There is a forthcoming chapter on the topic written by the specialists Manuela Cantinho Pereira and Miguel Telles Antunes which will lay out the state of the art on the matter. Unfortunately, however, the article is under embargo and no further research on the collections is encouraged, or allowed, before its publication.

[5] There was no mention of those objects on the correspondence between Joseph Banks and Correia da Serra that I consulted and that the authors of the scientific biography of Serra (Simões *et al.* 2006) systematically examined (Simões, Diogo and Carneiro, personal communication).

thing can be gathered regarding the *Hau* of those objects: they possess the magical ability of bringing new subjects to life.

Beyond British Columbia and Hawaii: Banks and Serra in England

Whether or not Banks and Serra, solo or as a team, ever touched, carried, studied, gave, purchased or even saw the Hawaiian and British Columbian objects that are now in the Academy of Sciences in Lisbon, the fact is that the two men indeed shared many interests, knew each other, corresponded, and even went on a joint scientific expedition to the Yorkshire coast. Serra may have been just one more of Banks many acquaintances and interlocutors, but everything indicates that he was an esteemed one (Simões *et al.* 2006).

Sir Joseph Banks had a vast network of correspondents and interlocutors. He was the naturalist of reference at the time, the person who most clearly epitomised the making of knowledge, exploring nature, collecting, naming, comparing, cultivating and foreseeing the uses of such knowledge (Gascoigne 1998). In 1766, at the age of 23, he was already a member of the Royal Society. His first great expedition, that same year, brought him to Newfoundland and Labrador on board of the vessel *Niger*. He described the species found there according to Linnaean terminology (Lysaght 1971). Since the ship stayed in Lisbon for six months, he became acquainted with the place and with some local scientists, with whom he formed lifelong relationships–such as with Domenico Vandelli, who would decades later ask Banks for protection in England during a period of political intrigue in Portugal.[6] But the Abbot Correia da Serra, then 15-16 years old, was not in Portugal at that time–he was probably in Naples with his family, pursuing studies he would continue in Rome. Banks and Correia would meet later (Teague 1997, Simões *et al.* 2006).

When Correia da Serra moved to England in a rush, in 1795, out of fear of persecution, he was already a mature man of science with good credentials and connections throughout Europe. He had been the co-founder of the Academy of Sciences in Lisbon, and reputedly its mastermind. He had lived in Rome, Naples and Paris in past years, and would live again in Paris and in Philadelphia in the future. He knew James Edward Smith, interacted with several other peers, like Richard Anthony Salisbury and Robert Brown (Diogo *et al.* 2011, 358). Banks was a most esteemed colleague, with whom he shared interests and engaged in joint

[6] Correspondence of Joseph Banks, Museum of Natural History, London.

work. While in London, Correia da Serra published extensively and participated in many scientific activities (Simões *et al.* 2006).

Correia da Serra's time in England was one episode in a life full of variety; his relationship with Banks was an important, cherished and prestigious one, but also one among many others. A worldly character with a wide range of interests, Correia da Serra drifted in different directions, to different places and towards different scientific interests. He hardly concentrated on a single topic long enough to write a monographic study but, instead, he made many small and incisive contributions in assorted fields. He travelled since he was a child; he moved easily, and made himself at home in different places, sometimes running from one place to the other in fear of persecutions. His was a very full and multifaceted life, elusive to his biographers up until now (Teague 1997, 11).

Who was this man who so uniquely epitomises the Enlightenment in Portugal and in Europe, who spent most of his life abroad yet had a crucial influence on his country, who embodies the character of the *estrangeirado*,[7] who co-founded the Academy of Sciences, who had important Masonic connections, who interacted closely with the major figures within and outside of Portugal, who made a home in Naples, Rome, London, Paris and Philadelphia, who came to be a dear friend of President Thomas Jefferson in the United States, to the point of having his own room at Monticello, Virginia, who took orders and earned the title of *Abbé* (Abbott) but never practiced much, who had a common law wife and a son, who befriended people from the ancient regime and the revolutionaries that fought against it? The variety of characters he embodied mirrors the complexity of interconnected worlds that coexisted at that time of radical change in politics and in knowledge-making. He was a man of his time, he resembled the time he grew up in; he engaged in different streams of knowledge—letters, philosophy, natural history, botany, carpology (study of fruits), geology, ethnography, politics, diplomacy (Diogo *et al.* 2001). He was a polymath, in the words of his biographers (Teague 1997; Beale Davis 1993).

---

[7] Literally, "foreignised", or those who have undergone foreign influence. Carneiro, Simões and Diogo (Carneiro *et al.* 2000, Digo *et al.* 2001) suggest that the best translation is "Europeanised". Mostly as an effect of political exile from authoritarian regimes–Pombal, above all–the *estrangeirados* abounded in late eighteenth century Portugal and contributed to major reforms in education and science. While most of them merely imported to the country what had been developed elsewhere, Correia da Serra genuinely contributed to the advancement of knowledge on several fronts (Diogo *et al.* 2001, 354).

Correia da Serra, a polymath of the Enlightenment: his youth

José Francisco Correia da Serra came into the world in 1751 in the southern town of Serpa, Portugal. His maternal and paternal families had known each other and maintained a close relationship for generations, a solidarity that may have emerged from their status as New Christians—meaning that their great-great-grandparents had been forced to convert from Judaism to Christianity (Teague 1997, 29-30). Throughout the centuries they suffered Inquisitorial accusations of "Judaism, heresy and apostasy" (Teague 1997, 30). When José was ordained in Rome, he produced the required documents denying that his parents, grandparents or great grandparents were prosecuted by the Inquisition. It is unlikely that he was unaware of his ancestors' ordeals during the Inquisition; more likely, he was concealing his Jewish and freethinking background on purpose (Teague 1997, 31-32).

His own parents had moved to Rome in 1756-57 anticipating Inquisitorial prosecution, according to most biographers (Teague 1997, 33; Diogo *et al.* 2001, 355, note 2). In Rome, his father Luis Dias Correia kept a successful medical practice and trade in medicines. Yet they had to move again in 1760, when the Portuguese were expelled from the Papal States as retaliation against Pombal's expulsion of the Jesuits from Portugal. The family moved to Naples and lived there for the following decade. New births expanded the family, but several died in childhood. In 1765, his mother Francisca died at the age of 31, giving birth to her 12th child—a girl who became one of the five siblings, from a total of twelve, who made it to adulthood (Teague 1997, 27).

José lived in Naples from the ages of nine to 19 (Teague 1997, 37). It is known that while in Naples he studied, although briefly, with the abbot Antonio Genovesi (Diogo *et al.* 2001, 355); it is less clear whether it was there that he studied with Luis Antonio Verney, a reference figure of the Portuguese Enlightenment and the author of "The True Method of Study" (Verney 1746), since their periods in that city did not necessarily coincide (Teague 1997, 38); the two may have met in Rome (Diogo *et al.* 2001, 356). It is also likely that the boy was influenced by the vibrant Napolitan culture of that period, with its lively *palazzos,* semi-secret associations and temples of knowledge—among which the famous pharmacy of the *Ospedale del Reame (Incurabili)* stands out (Valerio 2010; Rispoli 2010).

The young José went on to study in Rome with the priest Maratta, who contributed to his taste for botany; he also took classes in medicine and frequented the Papal *Orto dei Simplicy.* At the age of twenty he corresponded with Carl Linnaeus and Antonio Turra (Teague 1997), although later in life he chose Jussieu's system over Linnaeu's (Diogo *et al.* 2001). While in Rome, José planned a scientific expedition to Sardinia with his college roommate Jean Desmete, a Frenchman. Instead, however, they went for a shorter trip to Liguria in 1774. On that trip he wrote abundant notes and reflections on everything he observed, natural or social and cultural, from geology and botany to diving and fishing methods (Correia da Serra 2003).

In the meantime, José was ordained. Whether due to family pressure, efforts to conceal his research activities from the Inquisition (Carvalho 1948), or merely the pragmatism of security and protection (Simões *et al.* 2006), he became a man of the church, the Abbé Corrêa da Serra, yet one who was not overly enthusiastic about the endeavour (Diogo *et al.* 2001; Teague 1997, 45). He engaged in activities often opposed by the church: science, research, secret societies, civic associations to promote the development of science. He had relationships with women and even had a son with his French partner Esther Lavigne. He engaged with the free thinkers that on the opposite side of the Atlantic built the United States of America. He also took part in scientific expeditions, although more modest than the round-the-world explorations of Cook and Banks, to the English coast of Yorkshire, the Italian coast of Liguria, the Portuguese island of Berlenga, and other sites in Europe and the United States.[8]

Correia da Serra, Lafões and the Royal Academy of Sciences

When Correia da Serra returned to Portugal in 1777, after his studies in Rome, he found he had just lost his father, who had returned home a few years before. Dr Luis Correia had lost his wife Francisca and his business had been in frank decline. José Francisco Correia da Serra inherited no wealth, but stress and sorrows—which would erupt to torment him later in life.

[8] His travel papers include: "Journal d'une Course en Avril 1774: Avec Mr. Demeste"; "Observations d'histoire Naturelle dans le Patrimoine de S. Pierre en 1776: Juin avec Mr. L'Abbé Chaupuy (?)"; "Observations faites en parcourant l'Espagne et le Portugal: N° 1. Voyage de Cadiz a Serpa, 1777"; "Observações sobre a formação e estructura Fisica das tres Provincias Meridionaes do nosso Reino"; "Observações feitas a huma jornada pela Provincia do Alentejo em Mayo e Junho de 1785"; "On a Submarine Forest, on the East Coast of England"; and "Observations and Conjectures on the Formation and Nature of the Soil of Kentucky", compiled in Correia da Serra (2003).

To his fortune however, the rich and influential Duke of Lafões had also returned to Portugal following a long exile during Pombal's government under King José. Lafões was a close relative of the new Queen, D. Maria, enthroned in 1777. He was also an opponent of the ancient regime, a sympathiser of the ideals of the Freemasonry and an enthusiast of progress, science and human growth through learning.

Through some connections with Beja's friar Manuel do Cenáculo, a man close to Pombal's circle of influence and at some point the head censor of the country (Carneiro *et al.* 2000, 599), Correia da Serra accepted the sponsorship of the Duke of Lafões and in 1779 moved to his palace in Lisbon–the *Quinta dos Alfinetes ao Grilo,* in today's Beato-Xabregas neighbourhood. There he found the material, emotional and social support necessary for his intellectual work. There he wrote in abundance; he left behind many of his notes when he fled to England, even though Lafões later denied their existence in order to cover up for his friend. It was also there that he and Lafões planned the details for the society they were about to create. Its purpose was to promote science, research, knowledge and freethinking in general, to push the country forward, away from the ancient regime's lethargy and backwardness that they abhorred so much. That society would become the Royal Academy of Sciences—"the sole sheet anchor in this wreck", as Correia referred to it in a letter to Manuel do Cenáculo (quoted by Diogo *et al.* 2001, 356).

The Royal Academy of Sciences was officially founded on December 24, 1779, sponsored by the Queen D. Maria. The main figureheads were the Duke of Lafões, its first president, and Correia da Serra, who became its first secretary and was appointed permanent secretary in 1788 (Diogo *et al.* 2001, 357). Soon botanist Domenico Vandelli joined them. Although the latter pushed for a focus on natural science, the Academy kept broader interests. With the motto "*NISI UTILE EST QUOD FACIMUS STULTA EST GLORIA*", the Academy bylaws stated its commitment to promoting public happiness, developing national education, improving science and arts, and advancing popular industry (see Cardoso 1990, Figueiredo forthcoming).

The Academy had different sections, or classes: natural sciences, hard sciences, beaux-arts and letters. At the time, letters or *Literário* meant more than fiction, and included most written work involving knowledge. There were full members, corresponding members, and foreign members. Via personal networks, international celebrities were invited to become corresponding or honorary members of the Academy and earn yet another distinguished title while broadening the Academy's networks, reach and prestige (see Simões *et al.* 2006, 53-54). Joseph Banks was among these, along with other key scientists and people of knowledge of the time. As noted by today's analysts, "in order to be effective, the Academy should function both as a centre of production and dissemination of knowledge in the country and as a part of an international network of scientific institutions" (Diogo *et al.* 2001, 356-7). If we follow the analytical suggestions of Actor-Network Theory (e.g. Law 1992), networks per se could account for the Academy's existence and functions. However, the genuine commitment of some of its members to progress, science, knowledge and human improvement should not be left out of the analysis.

In spite of some difficulties, the Academy was able to maintain an independent agenda and worked in parallel with the state. It attempted to keep the production of knowledge in constant flow while the governments drifted and shifted. The Academy, not the government, defined the methods of data collection in overseas territories and provided the guidelines for surveying the mineral, vegetable and animal worlds of each place, as well as the notes to take on the local societies. This created a core of organising principles for specimen collection and for the questionnaires that colonial officers in Africa and Asia used as a basis for their reports. Occasionally, the Academy promoted special expeditions, such as that of Alexandre Rodrigues Ferreira to the Amazon (Raminelli 2008).

In years to come, the Academy was either supported or targeted by the governments, but it always survived with an independent agenda. The headquarters moved between different temporary locations until they settled in the former Convent of Jesus, in 1834, where they remain today. The street was renamed *Rua da Academia das Ciências*. The reform of 1851 divided activities into two sections—Letters and Sciences, much similar to what today would be called Humanities and Sciences.

The Academy supported public education not just in theory but also literally. Its headquarters temporarily hosted higher education and entire sections of the University of Lisbon in the early twentieth century. Currently, the headquarters house the geology museum, which is open to the public, and the not yet open Maynense museum (after Father Mayne's class), with a precious collection that includes the Hawaiian and Pacific pieces mentioned earlier among other ethnographic rarities, as well as zoological and botanical specimens, scientific instruments and the remains of the former convent. Besides a vast library and numerous paintings and sculptures, the Academy produces its own publications, including the *Memorias da Academia*. The building itself is a gem with a majestic, gilded main hall, marble-paved corridors and stairways, beautiful rooms and cloisters.

With the mission of promoting knowledge against the country's backwardness and its absolutist governments, politics has permeated the existence of the Academy since its very beginnings.[9] It had to compete with projects promoted by other groups, which had ties to former or future scientific societies, vying with networks for influence, fighting for the prevalence of the knowledge produced according to the principles of science and reason. There were also some periods of lethargy; when Link visited Portugal at the end of the eighteenth century for the botanical survey that led to his monumental Flore Portugaise (Hoffmansegg and Link 1808-20), he found an incipient scientific society that had little to be proud of (Link 1801; see also Bastos 1988, Diogo *et al.* 2001, Simões *et al.* 2006, 73).

Those were times of intense turmoil, interrupted projects, short term governments and shifting political agendas. The waves of republicanism that came from France and its sympathisers were felt as a threat to the established powers; scientific activity was often considered suspect and scientists were accused of Jacobinism and Freemasonry, and the Academy was seen as a breeding ground for these ideologies (Taegue 1997, 57). Both Correia da Serra and Lafões were constantly targeted by the police superintendent Pina Manique (Diogo *et al.* 2001, 357). For reasons not fully documented, Correia da Serra moved to France in 1786 and returned only in 1791–right before and after the French revolution. In 1794, the French physician of Girondist persuasion Pierre Maria Auguste Broussonet, a known freemason, came to Lisbon and was hosted by the Academy of Sciences. Pina Manique was after him, and the pressure over Correia da Serra increased. After Broussonet escaped, so did the Abbé, who sought refuge in England in 1795(Diogo *et al.* 2006, 59-61).[10]

---

[9] The relationship between the Academy and the Portuguese state is intriguing up to our days. No longer a private association but under the sponsorship of the government (Ministry of Education and Science), its collections are kept under reserve and only occasionally become available to the public–as if the Academy's original mission of promoting science above the state was still in the agenda.

[10] Diogo *et al.* (2006, 60, note 10) transcribe a document from the Inquisition accusing Correia da Serra of a multitude of crimes and misdeeds of sexual nature–naming several women and a young man, plus referring to a number of unknown victims.

The wide networks of the enlightenment: Correia da Serra and Leonor de Almeida, Countess of Oeynhausen, Marquess of Alorna

After moving to London in 1795, Correia da Serra interacted with prominent English naturalists like Joseph Banks, James Edward Smith, Richard Anthony Salisbury and Robert Brown (Diogo *et al.* 2006, 77). Although the *Abbé* found shelter and support for his work in science, he could not secure enough income for his needs. Throughout his life he had a steady stipend from the house of Lafões (Monteiro e Costa 2006), but occasionally he encountered difficulties in making ends meet (Diogo *et al.* 2006, 63). He was chosen for a position at the embassy, but political intrigue prevented him from achieving it; the embassy was in the hands of the Marquis of Ponte de Lima, whose connections with the Inquisition revived the accusations against Correia da Serra (Diogo *et al.* 2006, 74). Seen from today's perspective, Correia's attitude may be considered slightly paranoid, as he so often feared persecutions that did not take place–yet, the accusations existed and supported his feelings of vulnerability, while fuelling his tendency to move from one place to the next.

In 1801, he moved to Paris where he interacted with the major naturalists and philosophers of the time; Lafayette, Cuvier, Pyramus de Candolle, Alexander Humboldt, Christian Persoon, Du Pont de Nemours, Julien La Metrie, plus the Portuguese botanist in exile Avelar Brotero (Carneiro *et al.* 2000, 609; Cardoso 2013, 19). For years, Correia da Serra had a fulfilling life. He had a relationship with Esther Lavigne, with whom he had a son, despite his vows of celibacy as a catholic clergyman. After some years however, France became uncomfortable for his demanding persona. What had once been the place of free thought, and thus the ideal setting for the pursuit of knowledge, had turned into a land of authoritarianism. He refused to write the eulogy of Napoleon and once again packed his belongings and left (Diogo *et al.* 2006, 127).

Correia da Serra crossed the ocean in search of freedom in America. He arrived in Philadelphia in 1812 with letters of recommendation from distinguished French scholars (Carneiro *et al.* 2000, 608). He taught at the University of Pennsylvania, became a member of the American Philosophical Society, engaged in several scientific expeditions and became close friends with Thomas Jefferson. Correia da Serra was a regular guest at Monticello, Jefferson's place in Virginia, where the "Abbé's room" can still be visited today (Teague 1997, Diogo *et al.* 2006, 132). Recent works on Correia da Serra in America show how intertwined he was with American politics (Beale Davis 1993, Almeida 2013, Cardoso 2013).

But once again he planned to leave; he entertained plans of moving to Brazil, then a colony of Portugal—or, should one say, its centre, albeit a temporary one. In anticipation of the Napoleonic invasions, the old and now demented Queen D. Maria,[11] her son João (future João VI) and a vast number of courtiers sailed from Lisbon to Rio de Janeiro at the end of 1807, establishing it as the capital of the empire in 1808, only to return to Portugal in 1821 (Wilcken 2006). Politics changed dramatically in 1822, when Brazil became a new independent nation and the liberals won in Portugal.

After years of incertitude and occupation by both Napoleon's troops and the English who had come to the country to fight the French, Lisbon was again in the hands of the Portuguese. And to Lisbon it was that Correia da Serra moved as well in 1822. No one can say whether this restless character would be ready to move again in a few years' time; his life was cut short in the following year. He died in 1823 on a trip to the spa at Caldas da Rainha, where he sought treatment for the ailments he had accumulated throughout his life.

In each of the different places he lived, Correia da Serra expanded and cultivated an impressive number of friends, colleagues, interlocutors; a few of them were—like himself—well-travelled, cosmopolitan Portuguese men such as Luis Antonio Verney, his senior, and the Duke of Lafões, co-founder and mentor of the Royal Academy of Sciences. But there was also at least one well-traveled Portuguese woman he interacted with throughout his life: Leonor de Almeida (1750-1839), Countess of Oeynhausen and future Marquess of Alorna, herself a key element in the European Enlightenment. At least once, in 1785, Correia travelled with Leonor and two of her children, who were on their way from Vienna to Lisbon and took him on board in Avignon.[12] A prolific writer of letters and diaries,

[11] The loss of her eldest son, José, who should have become the next monarch (José II), devastated the queen to the point of no return. It is still unclear whether his death was due to accident, poisoning or smallpox. Wilcken suggests smallpox and relates the queen's insanity to plausible feelings of guilt, for she had chosen not to inoculate him at a time when such preventive action carried many dangers (Wilcken 2006).

[12] In a unique book restoring the combination of research and poetry, Maria Teresa Horta presents her extensive research on the life of Leonor de Almeida in the 1,000 pages literary masterpiece *As Luzes de Leonor* (Horta 2011). I am immensely grateful to the author for the book and generous sharing of her research in subsequent encounters.

Leonor portrayed him as great company, delicate and cultivated, full of humour, although a hypochondriac (Horta 2011, 531), and above all an insatiable seeker of knowledge, using every halt of the carriage to go into the bushes and collect plants, which he later dried, studied, compared and named (Horta 2011, 539-540).

Who was this unique person who travelled almost as much as Correia and, like him, was at home in many different places, and whom, unlike him, belonged to the upper aristocracy, and befriended both the old French aristocracy and the revolutionaries whose ideas nourished her freethinking mind?

Leonor de Almeida was born into the family of the Marquis of Távora, whose members suffered major persecution by Pombal and were killed or jailed without any criminal evidence. Leonor grew up imprisoned in a convent in Chelas, Lisbon, with her mother and sister, while her father and brother were held captive elsewhere and her grandparents were tortured to death. Although in jail, she grew up to be a reader and a writer–her literary name would be *Alcipe*. Forbidden books of science, philosophy and poetry entered the convent in disguised ways; she read Voltaire, Rousseau and the encyclopedists. At the *grade*, the iron grate separating the imprisoned from the free, there were weekly meetings where she could interact with visiting poets and literati who came from all over the country. They mostly came to hear her readings and her poetry. Among them there was another woman poet, Teresa de Mello Breyner who is likely to have helped distribute Leonor's writings in outer circles.

Freed at the age of 27, after the end of Pombal's dictatorship and under the new political regime of the Queen D. Maria, Leonor pursued a life that combined her aristocratic sociabilities and her passion for knowledge and poetry. After her marriage to the count of Oeynhausen in 1779, she moved to Vienna; like the Duke of Lafões, she was exposed to the aristocratic version of the Enlightenment practiced in Vienna, different from the more bourgeois French agenda. She returned many times to Portugal, where she was close to the Queen—around whom, however, political intrigue abounded. Not merely a "Portuguese Mme. Stael", as she is often referred to, Leonor made a unique contribution to the development of knowledge and went about life in inimitable ways (Anastácio 2009; Horta 2011).

Beyond the gentlemen's clubs: a gendered account of the Enlightenment and its scientific societies

Leonor had an important influence in the development of Portuguese letters and knowledge in general; she is acknowledged in history textbooks as one of the *estrangeirados* who brought in novelties from abroad and who herself contributed to culture via her poetry. Although most historical literature from that period refers to *Iluminismo* (the Italianised version for having light, for enlightenment), she preferred the term *Ilustração*, which might better be translated as cultivating, learning (Bello Vazquez 2005, 47-52, Horta 2011). Either way, she was a woman of the Enlightenment, whose life and work have been at the centre of numerous writings (e.g. Anastácio 2009; Horta 2011).

The study of Leonor de Almeida's life and accomplishments provides us with a counter-narrative to the repetitively male-only accounts of the Enlightenment and the historical scientific societies. Acknowledging her contribution implies a revision of the gendered understanding of the Enlightenment networks; it also leads us to a number of other women who played a role in the literary societies of that time and who have remained until recently almost invisible (Bello Vazquez 2005; see also Bombelles 1979; Monteiro and Costa 2006).

As recent scholarship has shown, there were several women who actively engaged in literature and in promoting knowledge in Portugal at the time; Bello Vazquez (2005) lists the Viscountess of Balsemão (*Natércia*), Joana Rousseau deVilleneuve, Leonor da Fonseca Pimentel, Soror Theresa Angelica Peregrina de Jesus, the Countess of Pombeiro, the Countess of Soure, Joanna Isabel de Lencastre Forjaz (*Aonia*), Maria do Patrocinio, Maria Lobo, a nun Maria do Monte, Francisca de Paula Possolo da Costa (*Francilia*), Ana Bernardina Pinto Pereira de Sousa Noronha, Rita Clara Freire de Andrade, Ana Josefa de Bivar, Isabel Peregrina de Salazar Vasconcelos e Crato, Maria da Graça Fortunata, Margarida Gertrudes de Jesus, Maria Antonia de S. Boaventura e Menezes, Maria Teresa Margarida da Silva e Orta, Catarina Micaela de Sousa Cesar e Lencastre, Micaela dos Prazeres, Angela do Amaral Rangel, some of them frequenters of Leonor's poetry circle.

Most notable among them was the self-effacing Teresa de Mello Breyner, the Countess of Vimieiro (1739-18??), ignored by most historians of literature, sometimes misnamed as Vimioso and relegated to an insignificant role (see Bello Vazquez 2005). Like Leonor, Teresa was a literary writer of many talents. But what flamboyance Leonor possessed, Teresa had equally in discretion and modesty. She used the pen name of *Tirce* or *Tirse* (Bello Vazquez 2005; Horta 2011). But she went without using it as well. When in 1788 the Abbé Correia da Serra read the ticket that should contain the name of the secret author of *Osnía,* the literary work that won the Royal Academy of Sciences award, he found only a note saying that the author did not want to be known, and that they should use the award money "to reward the one who could indicate the best method to cure the olive trees that at the moment were ailing with a new disease."[13] The author was Teresa, but she remained anonymous for a long time. For many years, historians of literature attributed the work to a male author (Anastácio 2002; Bello Vazquez 2005).

A descendent of cultivated Austrian women on her mother's side–her grandmother Barbara Breunner was an attaché to the Austrian-born Maria Ana, wife of King D. José—and of the prominent house of Ficalho on her father's, Teresa had remarkable social capital, in addition to an education that set her apart (Bello Vazquez 2005). She married into the powerful house of Vimieiro, equally given to learning and education. Rich, influential, learned and talented, she may have been the shadow figure of the Royal Academy of Sciences, doing the basic work upon which everything rested, yet modestly erasing the traces of her presence (Anastácio 2002, Bello Vazquez 2005, Horta 2011). After her husband's death, Teresa chose to move to a convent interrupting most of her social activities and the date of her death remains unknown. Much is still to be known; the work of Raquel Bello Vazquez on Teresa de Melo Breyner (2004, 2005) and the circle of women and men who not only contributed to science and learning but also influenced politics is just the beginning of a trend that may lead to a novel understanding of the gendered politics of scientific societies of the past.

Concluding note

We began at the present day Academy of Sciences with one of its collections, an intriguing set of objects layered with references to the past, distant places, and transnational networks that all contributed to their reaching the Academy's cases. Whether the objects indeed circulated via those networks is unimportant: they stand as a metaphor for the insertion of the Academy of Sciences and its members in the networks of the Enlightenment, via which people and ideas circulated. By closely studying the circulation of some of those people and their role in the Enlightenment we reach a scenario that calls for a gendered appreciation of the Enlightenment and one of its icons in Portugal, the Academy of Sciences.

---

[13] "Ao abrir o bilhete, que devia conter o nome do Author da Memoria coroada, se achou que elle se nao dava a conhecer, e so desejava que o premio fosse applicado a quem indicasse o melhor methodo de curar o mal que actualmente ataca as oliveiras em varios sitios deste Reino" (Gazeta, trancribed by Bello Vazquez 2005, 88).

Acknowledgements

This chapter results from research conducted within the project *SOCSCI Scientific societies in contemporary science* (PTDC/CS-ECS/101592/2008), under the coordination of Ana Delicado at the Institute of Social Sciences, University of Lisbon. I thank Ana Delicado and all the team members for the simulating intellectual exchanges, and I cherish the privilege of supervising research assistant Patrick Figueiredo in the historical component of the collective project. The first version of our reflections on the relationship between Banks and Serra was explored in the paper "Enlightened networks: Joseph Banks, Correia da Serra and the development of the Royal Academy of Sciences in Lisbon", presented by Bastos and Figueiredo at the conference *Exploring Empire: Sir Joseph Banks, India and the 'Great Pacific Ocean': Science Travel, Trade, Literature, Culture*, National Maritime Museum, Greenwich, 24-25 June 2011. After the presentation I was able to engage on a brief reading of Banks correspondence at the Museum of Natural History in London. Further reflections on the networks of the Enlightenment were possible on a visit to Naples in September 2012 under the project *The science, clinic and art of syphilis* (FCT/HC/0071/2009), coordinated by Cristiana Bastos at the Institute of Social Sciences. Rich discussions about circulating objects and knowledge networks arose in the context of the project *Colonial Mimesis in Lusophone Asia and Africa* (PTDC/CS-AN T/101064/2008), coordinated by Ricardo Roque at the Institute of Social Sciences, of which this text is also a result. I am thankful to all my colleagues and interlocutors for their contributions to these reflections, in particular to Nuno Monteiro for sharing his knowledge on Lafões and his entourage, and to Maria Teresa Horta for sharing her insights on the network of women and men that gave life to the Royal Academy of Sciences.

Bibliography

Almeida, O. T. 2013. "Prefácio". In R. Beale Davis, *O Abade Correia da Serra na América 1812-1820*. Lisbon: Imprensa de Ciências Sociais, 9-12.

Anastácio, V. 2002. "Mulheres varonis e interesses domésticos" Reflexões acerca do discurso produzido pela História Literária acerca das mulheres escritoras da viragem do século XVIII para o século XIX. *Actas do colóquio Literatura e história*. Lisbon: Universidade Lusófona de Humanidades e Tecnologias, 427-445.

—. 2009. *Marquesa de Alorna (1750-1839). Estudos.* Lisbon: Prefácio Editores.

Bastos, C. 1985. "Literatura, botânica, sistemática e ciências sociais: a viagem de Link a Portugal". *Arquivo de Anatomia e Antropologia* 39: 173-180.

Beale Davis, R. 2013. *O Abade Correia da Serra na América 1812-1820.* Lisbon: Imprensa de Ciências Sociais.

Bello Vázquez, R. 2004. "Dá uma risada quando ouvires..." Transgressão e ocultamento em Teresa de Mello Breyner". In A. M. da C. Toscano, & S. Godsland (eds.) 2004. *Mulheres Más: Percepção e Representações da Mulher* Vol. I. Porto: Universidade Fernando Pessoa.

—. 2005. *Uma certa ambição de glória: Trajectória, redes e estratégias de Teresa de Mello Breyner nos campos intelectual e do poder em Portugal (1770-1798).* Universidade de Santiago de Compostela, Faculdade de Filologia.

Breyner, T. J. de M. 1788. *Osmía.* Lisbon: Academia de Ciências.

Brigola, J. C. 2003. *Colecções, gabinetes e museus em Portugal no século XVIII.* Lisbon: Fundação Calouste Gulbenkian

Bombelles, M. de. 1979. *Journal d'un Ambassadeur de France au Portugal 1786-1788.* Paris: Presses Universitaires de France.

Cantinho Pereira, M. M. 2005. *O museu etnográfico da Sociedade de Geografia de Lisboa: modernidade, colonização e alteridade* Lisbon: Fundação Calouste Gulbenkian.

Cardoso, J. L. (ed). 1990. *Memórias Económicas da Real Academia das Ciências de Lisboa, para o Adiantamento da Agricultura, das Artes e da Industria em Portugal, e as suas Conquistas.* Lisbon: Banco de Portugal.

—. 2013. "Introdução: revisitar Correia da Serra". In R. Beale Davis, *O Abade Correia da Serra na América 1812-1820.* Lisbon: Imprensa de Ciências Sociais, 13-21.

Carneiro, A., A. Simões & M. P. Diogo. 2000. "Enlightenment Science in Portugal: The Estrangeirados and Their Communication Source". *Social Studies of Science.* 30(4): 591-619.

Carvalho, A. S. 1948. "O Abade Correia da Serra". *Memórias da Academia das Ciências de Lisboa*, Classe de Ciências, sep 6.

Correia da Serra, J. 2003. *Itinerários histórico naturais.* Edited with an introduction by A. Simões, A. Carneiro & M. P. Diogo. Porto: Porto Editora.

Delicado, A. 2009. *A Musealização da Ciência em Portugal.* Lisbon: Fundação Calouste Gulbenkian.

Desmond, R. 1995. *Kew: The history of the Royal Botanic Gardens*. Kew: Harvill Press and the Royal Botanic Gardens

Diogo, M. P., A. Simões & A. Carneiro. 2001. "The Portuguese naturalist Correia da Serra (1751–1823) and his impact on early nineteenth-century botany". *Journal of the History of Biology* 34 (2): 353–393.

Gascoigne, J. 1998. *Joseph Banks, the British State and the Uses of Science in the Age of Revolution.* Cambridge: Cambridge University Press.

Figueiredo, P. n.d., "Da fundação aos primeiros anos da Academia das Ciências de Lisboa", submitted to *Análise Social.*

Hoffmannsegg, J. C. & H. F. Link. 1809-1820 (-1840). *Flore Portugaise ou description de toutes les plantes qui croissent naturellement en Portugal.* Berlin: Charles Fréderic Amelang.

Horta, M. T. 2011. *As Luzes de Leonor.* Lisbon: Dom Quixote.

Law, J. 1992. "Notes on the Theory of the Actor-Network: Ordering, Strategy and Heterogeneity". *Systems Practice* 5: 379-393.

Link, H. F. 1801. *Travels in Portugal: And Through France and Spain.* London: Longman and O.Rees.

Lysaght. A. M. 1971. *Joseph Banks in Newfoundland and Labrador, 1766.*

Mauss, M. 1925. Essai sur le don. Forme et raison de l'échange dans les sociétés archaïques. *L'Année Sociologique;* (included in *Sociologie et Anthropologie*, PUF, 1950)

Mauss, M. 1990 *The Gift: Forms and functions of Exchange in archaic societies.* London: Routledge.

Monteiro, N. G. 1998. *O Crepúsculo dos Grandes: A Casa e o Património da Aristocracia em Portugal (1750-1850).* Lisbon: Imprensa Nacional

Monteiro, N. G. & F. D. Costa. 2006. *D. João Carlos de Bragança, 2º Duque de Lafões*. Lisbon: Inapa

Obeyesekere, G. 1997. *The Apotheosis of Captain Cook: European Mythmaking in the Pacific.* Princeton: Princeton University Press. "With new preface and afterword replying to criticism from Sahlins."

Raminelli, R. 2008. *Viagens Ultramarinas: Monarcas, Vassalos e Governo a Distância*. Lisbon: Alameda.

Rispoli, G. 2010. *L'Ospedale del Reame. Gli Incurabili di Napoli, Vol II: Documenti i Riflessioni Scientifiche*. Napoli: Il Torchio della Regina.

Sahlins, M. D. (1985). *Islands of history.* Chicago: University of Chicago Press.

—. (1995). *How "Natives" Think: About Captain Cook, for example.* Chicago: University of Chicago Press

Simões, A., A. Carneiro & M. P. Diogo. 1999. "Constructing Knowledge: Eighteenth-Century Portugal and the New Sciences", *Archimedes* 2: 1-40.

Simões, A., M. P. Diogo & A. Carneiro. 2006. *Cidadão do mundo: Uma biografia científica do Abade Correia da Serra*. Porto: Porto Editora.

Teague, M. (selection and analytical introductory texts). 1997. *Abade José Correia da Serra. Documentos do seu Arquivo (1751-1795). Catálogo do Espólio*, Lisbon: FLAD.

Valerio, A. 2010. *L'Ospedale del Reame. Gli Incurabili di Napoli, Vol I: Storia e Arte*. Napoli: Il Torchio della Regina.

Verney, L. A. 1746 *Verdadeiro método de estudar para ser util à republica.* Valensa: Antonio Valle.

Wilcken, P. 2005. *Empire Adrift: The Portuguese Court in Rio de Janeiro, 1808-1821*. London: Bloomsbury Publishing.

CHAPTER TWO

THE ASSOCIATION OF PORTUGUESE ARCHAEOLOGISTS AND THE ESTABLISHMENT OF ARCHAEOLOGY IN CONTEMPORARY PORTUGAL

ANA CRISTINA MARTINS

European archaeology in the first half of nineteenth century

Since the Renaissance, scholars have cast their gaze on matters illustrative of a past that was intended to anchor Western civilisation. Therefore, preference was given to classical remains, transforming them into models to improve for the glory of the new age. However, the inextricable movements of the Reformation and the Counter-Reformation caused introspection on each locality, region and country. Specific substantive remains from pre- and post-Roman times were scrutinised in order to form from them identities outside the meshes of ancient imperiality. Gradually, the essential elements for the (re)construction of the past to consolidate political agendas built on the ideals of unity were brought together. The past was thus assumed as a means of ideological statement, regardless of its nature and extent. Not being anything new strategically, it was rooted in the turn of the eighteenth century, when the so-called nation-states were built, the logic of which had to be found in the deepest ancestry, connecting a territory with a way of living, being and doing, even if little or not at all aware of the effects of this conclusion.

Slowly, the European patchwork was weaved, cut up according to the ambitions, more or less disproportionate, of those who wished to shield themselves from surrounding realities. One of the first major zeniths of this approach was observed during the rediscovery of Herculaneum and Pompeii, in the shadow of Vesuvius, near Naples, metamorphosed by the future Charles III (1716-1788) of Spain, and VII of the Kingdom of Naples and the Two Sicilies (1816-1888), into a means of personal proclamation among the major European royal houses (Martins, 2011), or even that of D. João V (1689-1750) when he launched the 1721 Decree. Classical antiquity and the Greco-Roman aesthetic thus assumed increased power, strengthening a place that had never been denied to them since Humanism, despite cumulated diatribes caused by circumstantial displeasures. Neoclassical taste crossed all western borders, establishing itself among aristocrats and the bourgeoisie as a demand for social recognition comparable to their financial means. However, the end of the Enlightenment added a new curiosity, establishing itself forever in the western imagination and imagery. Indeed, the knowledge raised and translated to western publications and venues by experts incorporated into the Napoleonic expedition to Egypt (1798) profoundly modified the vision of the past, prompting the (re)assessment of transgenerational understanding sustained by the biblical record. However, at the same time as pre-classical and classical studies were being redoubled, Europe faced fresh unrest, led by Bonapartism, which was eager to encompass lands and peoples in the name of individual will. A foray oblivious to local desires, resulting in a substantial bolstering of hyperbolic nationalist demands for identity registered in different media and forms. Advancing into outmoded recesses, the Napoleonic Code instigated disruption anchored in unrepeatable uniqueness, as the only capable of structuring the cohesion essential to its survival. A context rather favourable, therefore, to the resumption of studies of national antiquities to highlight the classical iconography transposed to contemporary enticements, especially where the discourse surpassed most approaches. A framework exacerbated on British soil, where the continental blockade (1806-1814) reduced the number of persons undertaking the Grand Tour, which was replaced by wanderings throughout its multiple and diverse places, creating an archetype of action in the area of graphical recording of the monuments visited, regardless of their chronological and cultural attribution, often becoming the only indication of their existence.

The past was, then, completely incorporated into the political narrative, in particular by containing hegemonic programmes exalted by the Napoleonic conquests, accentuating nationalist reactions in an eager quest for the fundamental standardisation of the Industrial Age. It thus gave rise to Romanticism, born in the shadow of the ruins as a paradigm of transience and human *vanitas*. It also secured the romantic idealisation of ethnic and national differences, two inseparable realities hardly comprehensible when examined in isolation. The cult of the endogenous past and, by extension, its "antiquities", would therefore be led by a bourgeoisie frightened by the French invasions and an aristocracy stripped of its ancestral role. Throughout the world, romantics and liberals united in the name of preserving memories fundamental to (perceived) national identities, as opposed to the sayings and doings alien to their essences. Together, they promoted historical research, pursuing artistic studies along the Winckelmannian line and favouring a monumental and mobile inventory, fed by the Napoleonic wars, the British Ultimatum (1890) and the patronage of the neo-enlightened, such as D. Fernando II (1816-1885) and his descendants. This was an urgent procedure, following the dissolution of the Religious Orders in several European countries, such as Portugal (1834), plus the abandonment of property due to the exile and flight of owners vanquished in ideological quarrels in the wake of the collapse of the *ancien régime*. This situation was favourable to the emergence of organisations devoted to the study, preservation and dissemination of heritage, as a mainstay of contemporary western nurturing of some of its new and essential spaces of sociability and training, starting with museums.

In the meantime, something happened that shook the cultural and mental foundations of Europe. In 1859, in the academic world, the undisputed existence of pre-historicity was announced, given the association of remains of human bones, long-extinct animals and artefacts comparable to many others observable with the so-called "current primitive", in the same geological layer and in a closed context. Thus, a veritable Pandora's Box of unimaginable consequences for the daily lives of believers in the perfectible Demiurge was opened. Taken as *mirabilis* in the affirmation of prehistoric research, this *annus* was consolidated by the launch of the work that provided the structure for Darwinian theory, *The origins of species* (1859), originating, together, a broad discussion that has not disappeared altogether from the coeval annals, before nourishing the most heated of debates, justifying thematic exhibitions to the delight of those upset with predefined parameters. Initiatives in this direction therefore unfolded, specialist societies and journals were founded, exhibitions were organised and the subject was incorporated into university curricula. With natural impact on western imagination and thought, prompting caricatural censure and comments, pre-historicity entered the Portuguese world with some ease and without major torment, judging by journalistic articles and monographs, most of which were written by intellectuals trained in Paris and London during the internal strife between the liberals and the absolutist monarchists, transposing to Portugal ideas and ideals seized upon abroad.

Meanwhile, the urgent task of recovering and advancing a country devastated by consecutive fatalities laid the foundations for a general plan of public works involving a thorough knowledge of the lands to be gutted in order to erect upon them structures indispensable to the contemporary industrialised world. Thus the founding (in 1857) of the Geological Services Commission (CSG) within the recently created Ministry of Public Works, Communications and Industry (see next chapter) was predictable, in which the precursors of our prehistoric research collaborated. It was also (or above all) a space dedicated, primarily, to the production of scientific knowledge. It converged, then, for the synchronous growth of sciences as relevant as the geological, pre-historic and anthropological, creating the first work published in Portugal in these areas and the formation of a museum as a consequence of the incursions of its leading members, quickly incorporated into the mineralogical section of the National Museum of Lisbon, i.e. the former Museum of Natural History of Ajuda, maintained at the Royal Academy of Sciences of Lisbon (see previous chapter) (1779) and belonging to the Polytechnic School (1837). Generally, the transition from the eighteenth to the nineteenth century made evident an inner spirit more in accordance with the specificities of the natural sciences than the humanities, translated in the forms and content of collections and museums. Also, because of this, the uptake of pre-historic studies by amateurs and researchers was fast in terms of the characteristics of the subjects that consolidated them in the beginning: geology and palaeontology. Furthermore, work methodologies designed primarily for biology and adapted to their artefactual nature were applied: cataloguing and typological and geographical classification. Thus, the early attention paid by the West to elements of Natural History was expressed and reiterated, taking its own field of action among us, thanks, in part, to the Pombal reforms of the seventeen hundreds and speciality courses opened in the meantime (1836). At the same time, scientific research stimulated itself, attributing distinctive prizes and creating institutional links with major peer centres of scholarship.

This involvement allowed the definitive transfer of archaeology from a dilettante, amateur and antiquarian world to the scientific sphere. Describing archaeological discoveries in various corners of Europe, the major national journals likewise reported expeditions prepared by major Western powers to the territories mentioned in the Old Testament, in search of a long-lost Paradise and illustrations testifying to the most renowned episodes recorded in the Gospels. However, what is important is to emphasise that it was precisely after 1859 that, at the confluence of several approaches, working methods (taken from geology, zoology and anthropology) and theories such as Evolutionism and Natural Selection, a new specialisation emerged, prehistory, enabling human antiquity to be chronologically extended far beyond the dogmatic and reductive chronologies of the Bible.

More than archaeological studies as such, traditionally related to historical and artistic approaches, it is the pursuit of prehistoric man that deserved greater attention from the scientific community, both due to the commotion it involved and the holistic and interdisciplinary approach it imposed. Archaeology thus definitively came under academic jurisdiction and, more importantly, entered the university domain, the last stage of the public appreciation of its value, the social. By doing so, it opened the way to the (re)use of archaeology, in general, and prehistory, in particular, in agendas of a nationalist, regionalist and localist nature, if not with hegemonic and transnational elements. Therefore, the possibility loomed of researching and creating conditions essential to archaeological expansion, borne of a deep desire to break taboos and precepts forced for centuries by the powers that be, searching for the first roots of human reality/realities in an environment dominated by a return to the classical origin of Western culture.

Nineteenth century Archaeology in Portugal

Like the museums, in certain contexts, the emerging sciences played a key role in processes of political legitimacy. A proposition contradictory to the idea that it was precisely the social sciences and humanities that had the greatest difficulty in gaining acknowledgement, propagating and reaching the so-called "general public", especially by means of museums. The reasons were several, although the lack of three-dimensionality of the objects displayed was the most significant, distinguishing two emerging sciences in the nineteenth century, gaining ground on the others, while conquering terrain and drawing upon examples from the exact and natural sciences. At the confluence of the two main spheres of the production of knowledge–human and natural–archaeology and anthropology overcame the restricted world of the one- and two-dimensional narrative, thanks to their object of study: material culture.

Recognised by academia, archaeology needed to achieve political and public support as the driving force behind its scientific survival, especially through the establishment of specific societies, university departments and regular meetings regarding the speciality. This desideratum was also reachable thanks to the social position of archaeological adherents, devising relevance in the construction of "imagined communities", (between) dreamed and required. Only at the end nineteenth century did archaeology mature, fitting into the most scientifically and socially prestigious central institutions, the number of research projects, local and regional societies, publications and public presentations multiplying, and the first manuals being written. Clearly, the preparatory work for this gradual denouement began long before.

In addition to the first manifestations of sixteenth century curiosity and intellectual interest in the traces of antiquity on Portuguese soil, reinforced in the midst of enlightened absolutism and the natural philosophy movement, the winds of liberalism, the expansion of knowledge, the opening horizons and positivism focused gazes upon archaeology, in general, and prehistoric studies in particular. Meanwhile, the insistence on placing Portugal once again in the European game of political chess, regaining respect and esteem, justified cultural and scientific demonstrations to be compared with the best of what was produced beyond our borders. Looking at foreign examples, an association was organised for the study of a single archaeological site (of the Roman period), aspiring to enter the country into European itineraries with contemporary monographic seasons. Thus the Luzitana Archaeological Society (1849) was founded, whose name also clearly indicated the underlying ideas by invoking a claim to national ethnic, Lusitanian roots. However, its geographical position and permanent lack of financial resources prevented it from securing excavations, keeping up a newsletter and getting its own building to house a museum and library.

Awareness of the obstacles encountered by this forerunner; knowledge acquired as a student on French and Italian soil, especially regarding the central role of associations in the recovery of heritage; the sharing of the concerns of those who clamoured for the preservation of monuments; the understanding of how bureaucracy worked in Portugal; an extensive network of personal acquaintances, at home and abroad, all these facts allowed the architect of the Royal House, Joaquim Possidónio Narciso da Silva (1806-1896), to satisfy one of the mainstays of heritage awareness and, almost inherently, of affirmation of archaeological science among us. We may mention the foundation (1863/1872) of the Royal Association of Portuguese Civil Architects and Archaeologists (RAACAP) and its Carmo Archaeological Museum (MAC) (1864), converted into venues that defended, par excellence, archaeological activities, inspiring local scientific societies, some supported, with distinction, by its seconded members.

Indeed, this association managed to succeed where others would not, could not or did not have sufficient mental discipline and inner strength to consummate the initial project. A purpose that surpassed what was planned at the outset, the result of a series of circumstances, contexts and events to which it did not remain indifferent. On the contrary, it knew how to adapt to new historical, artistic and scientific guidelines in the face of a sense of emptiness in the spheres of knowledge, particularly in relation to pre-history, classical archaeology and art history. Therefore, from being an organisation craving the public recognition of the architectural profession, crucial in the conservation and restoration of monuments, the RAACAP quickly incorporated other perspectives regarding the past, extending its range of interests to impose itself on a relatively new area in the country and embrace new topics, research and members engaged in scrutinising the pages from bygone days in a refreshing and holistic manner.

However, something concrete worked as a catalyst for this new approach. What happened, then, between the time of its creation (1863) and the addition of archaeological activity, in the early 1870s? The answer lies in the presence of Possidónio da Silva at the Paris International Exposition of 1867. Moving once again to the city that had welcomed him during his higher education to attend the International Congress of Architects promoted during that event, Possidónio could not ignore the strength of the presence of prehistoric studies. From the outset, at the Trocadero itself, in its Galérie du Travail, where human evolution was presented from pre-historicity, based on artefacts in an unapologetically materialistic decomposition, that was (considered) more objective. After, at the Musée des Antiquités (later Nationales) (1867), opened for this very occasion, taking advantage of the presence of a significant number of adherents to the issues inherent in its collection. Finally, the holding, within the exposition itself, of the Second Session of the Congrès International d'Anthropologie et d'Archéologie Préhistoriques (CIAAP), led by Gabriel de Mortillet (1821-1898), one of the figures that most marked the destiny of European research in this new area of scholarship. Absorbing these modern directions intersecting with other thematic presentations, such as pre-classical Egypt and Mesopotamia, as well as all that promoted the safeguarding of monuments, so dear to French politics and regional interests, Possidónio returned to Lisbon committed to redirecting the association he conceived, utilising an entire network of pre-existing personal knowledge, and adding more that was essential to his new project. This strategy allowed him to add, in a visionary manner, the pursuit of archaeology to the association he ran, which became the central forum in the country in this area, assisting in the establishment of societies, studies and regional museums and sites through the intervention of corresponding members, at the same time organising free lectures and courses, launching the *Architecture and Archaeology Bulletin*–a model for other national journals–publishing compendia, collating data on the state of preservation of monuments and sites, and expanding the specialised library and eclectic collection of the MAC.

A succession of initiatives enriched with field investigations and prospects of the institutionalisation of archaeology and heritage protection, as in the case of the Ninth Session of the CIAAP, hosted in Lisbon in one year (1880) of great nationalist celebrations and excitement, such as the Camonianas (evocative of the sixteenth century Portuguese poet Camões, 1524-1580) and the reception that bordered on apotheosis of African explorers to throw into confusion European appetites for the territories of Angola and Mozambique. Realising its importance, especially due to the expected prestige, the Government hired Estácio da Veiga (1828-1891) to conduct an archaeological survey of the Algarve and select artefacts to be displayed in the cloister of the ARCL for analysis by the members of that Congress. The collection was packed into boxes, unfortunately, after the end of the meeting, until the doctor, philologist, ethnographer and archaeologist José Leite de Vasconcelos (1858-1941) incorporated them into the Muzeu Ethnographico Portuguez (1893). Contradicting the purpose of the Portuguese organisation CIAAP; the hopes fueled by the progress of the work; the public visibility afforded by the daily press to the issues discussed in the various sessions; the support of the Royal House for the event; the enthusiastic welcome given by the audience to the congress delegates, visiting archaeological sites properly prepared for this purpose, archaeology was not incorporated into an extensive plan to safeguard heritage, nor was it immediately welcomed in the universities, at least not explicitly.

After the congress, the Government asked the RAACAP to list ancient structures to be classified as national monuments, a public recognition of its work in this field. Delivered in a few months, thanks to the precious collaboration of corresponding members, the report grouped into classes, the items to be classified–Historical and commemorative standards of illustrious men–allocating one of them–the sixth and final–to prehistoric remains, "today, there being much appreciation for the study of Anthropology, they deserve *the government recommending to its officials* in their localities, *not to allow them to be destroyed*, as others were in ancient times." (1881, 14; our italics). It was hoped, therefore, to establish the pursuit of archaeology in Portugal through its introduction into consultative bodies and state executives and the university itself. It was, however, an illusion. Clearly, the efforts of the RAACAP were not completely inconsequential. Indeed, in 1881, the National Monuments Commission was created, although a permanent shortage of human and material resources prevented it from fully fulfilling its intended mission. However, instead of a body geared towards the archaeological area, Lisbon saw the inauguration of the National Museum of Fine Arts and Archaeology (1884), embodying a model that, to a certain extent, was already declining in the rest of Europe. Archaeology continued, therefore, outside the national political agenda.

Some disillusioned, others disgusted, it was left to those who were insistent and sustained in their convictions to raise their voices and implement long idealised plans, again in the private sphere. They met in learned regional and local societies, raised awareness, stimulated knowledge, attracted investment, and encouraged amateurs. However, the efforts were ephemeral, largely limited to voluntarism and to the betterment of a few. Even so, an attempt was made to address glaring gaps. Delaying the recognition of archaeology in university chairs, free courses disseminating archaeological, and specifically prehistoric, knowledge were advocated, such as the First Elementary course in archaeology, taught at the RAACAP (Martins, 2001). Similarly, venues, means and resources were modelled to focus attention and educate new audiences for the heritage crusade. The truth is that this commitment was not inglorious, as, albeit somewhat diluted, the University of Coimbra incorporated Archaeology in 1885, associating it, through prehistoric studies, with the chair of Anthropology and Palaeontology within the curriculum of the Faculty of Philosophy. Nevertheless, the death of some reiterated how much this area of knowledge owed to the private sector and to those who persisted in their demands, although not all possessed the resources to ensure its sound perpetuity, as in the case of the Martins Sarmento Society (1881). But the seed germinated with the indelible support of the RAACAP, despite conflicts between centralist and regionalist agendas, insurmountable due to the domestic political environment that, unlike other countries, dispensed with archaeology due to the lack of border, legal or identity issues nourished and substantiated by its studies.

The Association of Portuguese Archaeologists in the First Republic

Meanwhile, the dawn of the nineteen hundreds and the creation of a Republic established a new era in scientific consideration and the museum collections of some establishments. One example was the MAC, where, as in the MEP, there was a tendency to focus on an artistic theme, distinguishing aspects of "popular culture" (a source of national essence) by means of displaying pieces representative of its multiplicity, raising folklore in the understanding of European pre-historicity to the detriment of comparative ethnology, rooting the remaining manifestations of handicrafts in the deepest ancestry of each corner. Nothing to prevent, however, the redoubling of attention on archaeological research influenced by personalities such as J. Leite de Vasconcelos, who laid down the first major decree (at that time, still monarchical) classifying ancient structures as national monuments (June 23, 1910). Conceived in the Directorate-General for Public Works and Mines, the document included a group of "Prehistoric monuments", including "dolmens and other monuments", "Lusitanian and Lusitanian-Roman monuments" and "ruins", mirroring the contemporary interpretation of the past, archaeological theory and monumentality. Added to this was the creation of 13 regional museums of art and archaeology between 1912 and 1924, dependent on central power and sporadic municipal funding. Although not all incorporated archaeological collections, they expanded the word Republican in museological discourse, supported by museum resources, decentralising equipment, procedures and know-how in an initiative that formed part of a broad process of cultural resurgence envisioned by a Republic committed to raise its own profile. Emphasis was thus placed on local ethnography to reaffirm the nationalist programme of the new regime, which looked to merge anthropology and material culture, ethnogenesis and prehistoric artefacts. Furthermore, it was in prehistoric remains that the single trace of the past was sought, printed in the form and decorative grammar of the different typologies inscribed by ethnographic record, seeing in them their internal uniqueness and linear justification of the permanence of the same human communities in a space long delimited. Hence, one of the most prominent members of the (now) Association of Portuguese Archaeologists (AAP), the scholar of the culture of Lisbon, Gustavo Matos Sequeira (1880-1962) emphasised in 1922 the need to organise thematic exhibitions to strengthen municipal government, at a time when it had recovered politically, despite the (almost) ubiquitous centralist resistance.

It was in this broader context that the AAP renewed itself. With the death of its first proponents, the inflow of mainly archaeological and heritage advocates, and growing professional training, some of its architects left to found the Society of Portuguese Architects (1902). Added to the overthrow of the monarchy, this justified the redrafting of the articles of association, confining it to the archaeological sphere. In practice, however, the AAP continued to engage in eclectic themes of Portuguese history, interconnecting knowledge in a permanent exercise in interdisciplinarity that surprises due to the topicality of its procedures. This was translated into numerous conferences, thematic meetings, co-organisation of scientific meetings and exhibitions, and, last but not least, the *Association of Portuguese Archaeologist's Bulletin* (1913), and *Archaeology and History Bulletin* (1922), within a revealing hierarchy of the prevalence of the first studies over the latter, diluted until archaeologists began, largely, to direct their own destinies, without ever forgetting other areas of knowledge. However, increased attention was paid to local and regional affairs, resulting from an attempt to re-implement the spirit of the Arts & Crafts (1860-1910) to stimulate the national economy.

Not surprisingly, the AAP decided to celebrate its half-century anniversary with an authentic tribute to the past of Lisbon, its home city from the outset.

Thus arose the Lisbon Exhibition, commemorating the 50th anniversary of the association, inaugurated by the first President of Portugal Manuel de Arriaga (1840-1917), not in the correct year (1913) but in the following year, due to a succession of contingencies, the first of which was the delay in the works to renovate the interior of the MAC, initiated precisely to show the exhibits and welcome visitors with as much dignity as possible. Open to the public on March 24, the exhibition aroused great curiosity among the residents of Lisbon, who, possibly for the first time, had access to a set of rather disparate categories and typologies of objects illustrating the pages, more or less known, of the capital's past, emphasising the quality of traditional industries that were becoming renowned within and outside our geographical boundaries:

> [the speech] documents well what this event represented where the life of the capital became externalised, where its customs and traditions were easily manifested to the eyes of onlookers, where its monuments and their lost aspects could be noted again, where the Past spoke by means of the thousand eloquent and expressive voices of inanimate things (Archaeology and History Bulletin 1932, 90-91)

Moreover, the exhibition presupposed another purpose for the association. This was the need, for its own survival within the new political environment, to demonstrate scientific efficacy and a reorientation of action, without, however, denying some of its mainstays, reinforcing areas of study, such as genealogical and heraldic, involving certain circles less sympathetic to republican ideals. In fact, it was precisely from here that the AAP took on, more convincingly and boldly, a panoply of themes in a variety of specialised sections to jointly pursue paths and open up prospects consolidating its four future pillars: archaeology, history, studies of Lisbon and heraldry.

These objectives were achieved. If there were any doubts, the Lisbon City Council appointed an associative member to the standing committee for aesthetic evaluation of structures erected within its boundaries, referring to it often, until the Estado Novo (1933-1974) began to gradually distance itself from the decisions understood to be more relevant for effectively advertising an entire political discourse. However, additional proof of the prestige of the AAP was found when the municipality asked it to organise, at the MAC and under the leadership of its most prominent figures, an exhibition entitled City Museum (1922), a preliminary plan for one of the city's most interesting cultural facilities (1942). But that is another chapter in the history of the association that, in itself, deserves a separate text, to be published on another occasion.

Brief final words

The RAACAP / AAP metamorphosed into a unique forum for the discussion and dissemination of knowledge regarding safeguarding heritage and archaeology, defining and implementing plans for its recognition in Portugal so as to compare it to European realities, while seeking to adapt to the internal political changes, which justifies its 150 years of uninterrupted activity.

However, if this demand has managed to bring together experts and amateurs, supporting local initiatives and responding positively to government challenges, burying itself in field research, conferring, studying, doing museum work and publishing, the AAP has not gained its own place in Portuguese academia. Not as the AAP. It has, however indirectly achieved this through its leaders, many of whom occupy prominent places in organisations critical to the recovery of heritage and archaeological consolidation, whether in local associations or in city council committees or universities. Especially in the latter, where new members are recruited to perpetuate and innovate in line with best practices abroad. More than that, the AAP and, by extension, the MAC, has generated ideas, reoriented thoughts and actions, molded experts and forged closer links between university classes and archaeological fieldwork, providing, in the Bulletin, space to print its results.

Nonetheless, the path has been long, and sometimes tortuous. In fact, it quickly became apparent that it was a field of action feasible if incorporated into a greater plan of new ideas targeting the education of the public. Only by combating illiteracy and a scant aesthetic and cultural outlook can new audiences be won over, who will be expected to pursue the project initiated by the RAACAP, partially considered in the first

Republican legislation. It was in this way that an attempt was made to determine the consequences of the experience accumulated, as there are no single, perfect and inexhaustible models, especially in an area still so recent in Portugal.

Nevertheless, something failed, because the country was unprepared to understand, internalise, monitor and multiply a real experience firmly established in other European countries. Moreover, we suffer, in general, from a structural and structuring problem: the gap between the letter of the law and the essential means to comply with it. The absence of the required number of experts in the scientific areas connected to memory preservation and the pursuit of archaeology being one of them. Yet, the main reason for this (apparent) lack of interest in archaeology can be found in the absence of problems of internal and external assertion, deriving from a territory with borders defined long ago and no conflicts of a regionalist nature, especially those seeking autonomy, as witnessed in Spain. In short, archaeology was politically unattractive, inasmuch as it was ideologically expendable in a country that did not need to impose itself on others. It appears that archaeological research was supported (and is supported) only within agendas lacking local, regional or national affirmation by providing materials that assert its claims in an era of primacy of science and neo-rationalism. This distinction is also anchored in another aspect: the slow march of science among us and–more than that–the lack of a harmonious body of personalities that cultivate it, due to an inability to integrate into official structures, doing so mostly due to personal commitment. Furthermore,

> the rather faint-hearted action of a Central Committee for protection and surveillance, bereft of the means for effective action and justified existence, the systematic alienation of the municipalities, the indifference of parochial institutions, the absence of regionalist commissions to supervise and maintain local antiques, the general public disdain and intermittent publicity–by newspapers, magazines, books–of scholars, these are the already irreparable factors of the vast, noble and opulent national artistic/historical documentation! (Peixoto 1896, 330)

Even so, pre-historicity, for example, adjusted itself, to a certain degree, to republican ideals, highlighting the ancient human voyage around its own ability and desire. Nevertheless, it would not be sufficient to justify heavy investment in its systematic study, as it is unnecessary in the absence of issues of a nationalist nature. It was, however, enough for the AAP to press on with its purpose, remaining a model followed by various institutions that esteemed it from the outset in a continuous recognition of its efforts, the quality of its performance and excellence in the exercise of citizenship in this cultural and scientific field of activity.

Bibiography

Author unknown, 1932,"Discurso pronunciado pelo socio Gustavo de Matos Sequeira", *Boletim de Arqueologia e Historia*, vol. I, Lisbon, AAP: 90-91.

Author unknown, 1881, *Relatorio e mappas acerca dos edificios que devem ser classificados monumentos nacionaes apresentados ao governo pela Real Associação dos Architectos Civis e Archeologos Portuguezes em conformidade da portaria do Ministerio das Obras Publicas de 24 de Outubro de 1880*, Lisbon, Lallemant Frères.

Correia, V. 1916, "Arte Popular Portuguesa. Suas relações com a Arte Popular de toda a Europa", *Terra Portugueza*, 1 (3).

Delicado, A. 2009, *A musealização da ciência em Portugal*, Lisbon, FCG/FCT.

Diário Do Governo, I (124), Decreto 14, de 29 de Maio de 1911: 2244-2248.

—. (281), Decreto 1:700, de 18 de Dezembro de 1924: 697-703.

—. (34), Decreto 11:445, de 13 de Fevereiro de 1926: 135-146.

Martins, A.C. 1999, "O Muzeu Archeologico do Carmo e a Descentralização Cultural no Século XIX", *O Arqueólogo Português*, 4 (17).

—. 2001a, "Estudos Pré-históricos e Nacionalismo: uma Perspectiva Possidoniana", *Revista Portuguesa de Arqueologia*, 4 (1), Lisbon, IPA.

—. 2001b "O 1.º Curso Elementar de Archeologia (Lisboa, 1885)", *Trabalhos de Antropologia e Etnologia*, 55.

—. 2003a, "Património histórico-cultural: a emergência das reformas (do Liberalismo ao Republicanismo) 1.ª parte. *Património. Estudos*. 5. IPPAR: 31-50.

—. 2003b, *Possidónio da Silva (1806-1896) e o elogio da memória. Um percurso na arqueologia de oitocentos*. Lisbon: Associação dos Arqueólogos Portugueses.

—. 2004, "Património histórico-cultural: a emergência das reformas (do Liberalismo ao Republicanismo) 2.ª e última parte". *Património. Estudos*. 5. IPPAR: 109-125.

—. 2005a, "A Associação dos Arqueólogos Portugueses na senda da salvaguarda patrimonial. 100 anos de (trans)formação (1863-1963)." PhDThesis, University of Lisbon.

—. 2005b, "A memória da ruína, ou a ruína da memória?" In *Conservar para quê?*, ed. V. O. Jorge, Porto: Faculdade de Letras da

Universidade do Porto/ Centro de Estudos Arqueológicos das Universidades de Coimbra e Porto: 113-126.

—. 2005c, "Museu Arqueológico do Carmo: a Consagração da Memória", In *Construindo a Memória. As Colecções do Museu Arqueológico do Carmo*, coord. J. M. Arnaud and C. Varela Fernandes, Lisbon, AAP.

—. 2006, "O património, ou uma das fac(s)es da "Insustentável Leveza do Ser"?(!)". In *Cultura Light,* ed. V. O. Jorge, Porto: Faculdade de Letras da Universidade do Porto/Departamento de Ciências e Técnicas do Património: 35-52.

—. 2008, "(Instru)mentalizar saberes. A arqueologia como novo plano de poder." In *Crenças, Religiões, Poderes: dos indivíduos às sociabilidades*, eds. V. O. Jorge and J. M. Costa Macedo, Porto: Edições Afrontamento: 279-301.

—. 2011, "Olhares cruzados sobre a Cara Real portuguesa, a salvaguarda do património histórico-cultural e o exemplo bourbónico", *Lábaro* 2, Lisbon, Ordem Constantiniana, Delegação de Portugal: 15-37.

Peixoto, A. A. R. 1896, *O cruel e triste fado*, Figueira da Foz, Imprensa Lusitana.

Pereira, M. L. V. A. S. S. 1973, "O Museu Archeologico do Algarve (1880-1881). Subsídios para o estudo da Museologia em Portugal no séc. XIX". Thesis, Curso de Conservador de Museu, Lisbon, MNAA.

R. P., (1905-1908), "O Museu municipal de Bragança", Portugalia, II (1-4).

Sequeira, G. M. 1922, "Discurso". *Arqueologia e Historia*, I.

Veiga, E. 1880, *Memoria das Antiguidades de Mertola*, Lisbon.

CHAPTER THREE

RAILWAYS AND THE PORTUGUESE CIVIL ENGINEERS ASSOCIATION: PRACTICE AND THEORY (1870-1899)

HUGO PEREIRA

Introduction

In the first half of the nineteenth century, turmoil in Portuguese politics (feeble governments, frail majorities in the parliament, lack of political consensus) prevented the development of a public works investment programme, namely in the field of transportation (Mata and Valério 1993, 142).

In the 1840s, the minister Costa Cabral was able to introduce the necessary stability in order to decree the construction of the first railway in Portugal. Although the enterprise was unsuccessful, it became obvious that Portugal needed to invest in the modernisation of its transport and communication networks (quite archaic at the time, according to Guillemois 1995; Justino 1988-1989, 189-190; Matos 1980), so that the Portuguese could easily connect to Europe while exploiting the nation's resources (Santos 1884).

The coup of May 1, 1851, which initiated a period known in Portuguese History as the Regeneração (Regeneration), overthrew Costa Cabral, but not his economic strategy, which could be further implemented, since that revolution also brought pacification and an enlarged consensus to Portuguese political affairs (Bonifácio 1992, 96-98). In 1856 the very first Portuguese railroad was inaugurated. It was a small stretch of track with 36 kilometres between Lisbon and the village of Carregado. Until the end of the century, another 2,300 kilometres of tracks were added to that tiny railroad (Valério 2001, 372).

The state played a very important role in this matter, since every single aspect of railway construction and operation had to be approved by the government, namely by the minister of Public Works. Furthermore, a great part of the network was held and exploited by the state that also financed the majority of private enterprises (either with kilometric allowances or with guarantees of interest). In this process, engineers also played a leading role. Even though they might not have been the ones who decided which rail line was to be built (politicians in the government made that call), they would analyse the proposals, write down the contracts, survey the terrain prior to construction (often paid by the Portuguese state) and accompany the construction itself (Pereira 2012a, 359-403; Pinheiro 2008, 119-142). Until the late 1860s they did so in the advisory board of the Ministry of Public Works and in the Civil Engineering Corps of the same cabinet. From 1869 onwards, in response to budget cuts in the ministry, they gathered in the brand new Portuguese Civil Engineers Association, which they used as a means to lobby for the continuation of the public works programme.

In this chapter we aim to determine the role of this technical association in the development of the national railway strategy in Portugal. To do so we will analyse its opinions and compare those with the advices in the Ministry of Public Works, the discussions in the parliament and the actual decisions of the governments. The sources for this research will be the journal of the Association (the *Revista de Obras Públicas e Minas*), the manuscript reports of the engineers in the Ministry of Public Works, the parliamentary debates and the laws and decrees of the government. All of those will be examined under the scope of content analysis.

The creation of the Ministry of Public Works

A year after the 1851 coup d'état, the new government created the Ministry of Public Works, Commerce and Industry, where all matters related to railways should be dealt with. Within the new institution there was an advisory body (composed entirely by men with an engineering background) that could be heard by the government about issues related to public works and mines. However, not only was this hearing optional, but also its advice was not binding. The new ministry also published a monthly bulletin—the *Boletim do Ministério das Obras Públicas, Commercio e Industria* (The Ministry of Public Works, Commerce and Industry Bulletin)—that functioned as its official journal. This periodical, though, was more of a law compilation than an engineering journal. Among its contents we can find a few studies, surveys and statistical data, but most of its pages are filled with decrees, laws, ordinances and the like.

In sum, it did not work as a publication that could disseminate the theory and practice of engineering.

The advisory body (named General Council of Public Works and Mines until 1859 and afterwards merely General Council of Public Works) gave its opinion about a few railway projects presented to the government until 1868 (the Southern line between Barreiro and Vendas Novas in the Southern bank of the Tagus River; the Sintra line beginning in Lisbon; the Northern and Eastern railways connecting Lisbon to Oporto and to the frontier; and a few other proposals that were not implemented). In these occasions, its opinion was not fully respected a single time. In some instances, the government even decided to go against the better judgement of its advisory body. In others, the engineers in the Ministry of Public Works were not even queried. Such was the case when the government hired Sir Samuel Morton Peto to build the railway between Oporto and Lisbon in 1857. In this particular process (that lasted for two years) the General Council's judgement was never taken into consideration. Furthermore, according to the contract, every hypothetical dispute should be settled by the British Civil Engineering Association (Pereira 2012a, 161-163). By this time (and all the way through the nineteenth century for all that matters) the members of government and namely the members of the parliamentary Public Works commissions were the ones who decided which line should actually be built, regardless of the opinion of the General Council of Public Works. In addition, by that time, even though Portugal had some technicians, the governments preferred to rely on foreign engineers, who were regarded as more skilled and with a great deal of experience. The ministers would rather send some young Portuguese graduates to engineering schools abroad in order to acquire the necessary skills to one day take the public work programmes into their own hands (Matos and Diogo 2007). By the early 1860s, this was more likely to happen. However, in 1864, within the Ministry of Public Works is created the Engineering Corps composed of a few tens of engineers. By then, Portugal had inaugurated three major lines (Northern, Eastern and South and Southeastern) and, in addition, the government would decree, in 1867, the construction, based on Portuguese know-how, of two more lines in the Northern areas of the kingdom (Minho and Douro lines).[1]

[1] Collecção Official de Legislação Portuguesa 1864, 773-786; 1867, 441-442.

The reform of the ministry and the foundation of the Engineering Association

Nonetheless, the period was not suited for new investments. The companies that operated railways in Portugal (the Royal and the Southeastern of Portugal companies, of French and British capital, respectively) were going through severe financial difficulties and were pressing the government for help, lobbying in the financial markets against the issuing of Portuguese bonds (Pinheiro 1986; Vieira 1983). Amidst the Portuguese politics there was also a growing discomfort with the public works programme, in which Portugal had spent thousands of "contos" (one million réis, the national currency), gained very little and still had to compensate the losses of foreign investors. This situation led to the fall of the government and the empowerment of those who were more disappointed with the failure of the public works programme (1868). They belonged to the Reformist Party and reform they did. Railway construction was halted and budget cuts were imposed in the Ministry of Public Works. The Engineering Corps was extinguished, as well as the General Council (replaced by another body, the Junta Consultiva de Obras Públicas e Minas, whose members would earn lower salaries) and the ministry's bulletin. Fearing that this strategy could mean the end of the investment in public works and railways, the Portuguese engineers decided to form their own organisation that could defend their agenda and consequently lobby for the continuation of that strategy. Thus, in 1869, the Portuguese Civil Engineers Association was born (Brito, Heitor and Rollo 2003, 81-97; Diogo 1994; Pereira 2012a, 110-117; Rodrigues 1999, 76-82). Although many of these men did not share the political views and ideology of the statesmen in the government before the Reformist Party, they all agreed upon the necessity of investing in the modernisation of the national transport system. Railways meant progress and job opportunities, so everyone was happy. Hence, the new organisation became a privileged partner of the Ministry of Public Works and an authority as far as public engineering works were concerned. In addition, most of its members had strong connections to the state: some were public servants; others were councillors of the Ministry of Public Works; there were those who were congressmen or even ministers (Diogo 2003, 14; Matos and Diogo 2007).

The Association also edited a new journal entirely devoted to engineering—the *Revista de Obras Públicas e Minas* (Mines and Public Works Journal). It was supposed to fill in the gap left by the extinction of the *Boletim do Ministério*, but also to extend its scope. The *Revista de Obras Públicas* should announce the works done in Portugal and abroad and publish papers about engineering, something that the *Boletim* was never able to do. Its goal was quite clear: to create an archive of the undertakings and scientific and technological advances of the nation (Diogo 1994; Rodrigues 1999, 81-82).

Railways and the Association of Engineers

When the Association was created in 1869, Portugal had about 700 kilometres of railways in operation (Northern, Eastern, South and Southeastern lines) (Valério 2001, 372). It was not much. The Portuguese engineers in the Association were unanimous on the subject. Portugal had few railways and was well behind other European nations in this regard. Therefore, the country was in need of solutions and not theories. The *Revista de Obras Públicas* embodied this paradigm perfectly. In its pages practical papers and news about railways around the world should prevail over theoretical contributions (as one can read in the editorial of the very first issue of the journal).[2]

And in fact, in the following 30 years many were the contributions of the Portuguese engineers from the Association for the development of railways in the kingdom. Several issues of railroad building and operation were debated: roadbed, gauge, bridges, traction, rolling stock, predictions of income, etc. were some of the matters debated and divulged in the monthly journal of the Association. Some of these papers were written by engineers who were studying abroad in the engineering schools of Paris and Gand (Brito, Heitor and Rollo 2003, 25-30; Diogo 2003, 14; Matos and Diogo 2006, 5; Pinheiro 2008, 161-184). In this case, these articles served a double purpose: not only did they contributed to the debate of railway building in Portugal, but they also were a means to advertise the works of those engineers and to show just how tax payers' money was being used in the fellowships granted to those men. Furthermore, the *Revista* (and the Association itself, for all that matters) was a place where engineers from all over the country (each district had a fiscal director of public works, who was an engineer) could gather and exchange their experience and knowledge. This was particularly important to the growth of the knowledge of the country's topography. In the nineteenth century, the data about the orography of Portugal was very limited and scarce. The very first map of Portugal made with scientific basis dates from the 1860s. However, its level of detail was very low, since the scale was very large. In the following years, the information about Portugal's geography

---

[2] Revista de Obras Públicas e Minas, v. 1 (1870), n.1, 1-2.

increased due to the investments in public works (mainly railways and roads) and the work of the engineers spread all over the country. The engineers were the ones to conduct the studies of those works and survey the construction and operation of those railways (Branco 2003, 110-116; Dias 1995). Therefore, they had a particular good knowledge of a given area. They could then share that information with other colleagues and the *Revista* and the Association were privileged stages to do so. With a wider perception about the country's geography and economic needs and with a greater knowledge about the various details of railways building and operation, the Portuguese engineers were also able to survey, suggest and debate the construction of some individual lines.

The railway general plan discussion

Nonetheless, the input about the general railway plan was the greatest contribution of the Association and its journal to the railway debate in Portugal.

As we saw, when Portugal began setting its first rails there was no map with scientific basis where to draw a network. All the advices were therefore rather ethereal. As time went by, and as the engineers got to know the Portuguese geography better, the need to have a law decreeing the lines to be built in an organised and planned network grew stronger. At the beginning of the 1870s several entrepreneurs presented their proposals to the government to build a number of tramways or narrow gauge railways. By that time, the engineers both in the Ministry and in the association warned the government about the issues of conceding permits to build those lines haphazardly and without a general network plan.[3] The government heeded these suggestions and assigned the task of preparing that plan to the advisory body of the Ministry of Public Works (the Junta Consultiva) (Dinis 1915-1919, vol. 5, 336). This, however, was a poisoned chalice, as the task was overwhelmingly above the capacities of the Junta. The engineers in this institution were queried in most of the public works decisions taken by the government, be it the construction of a road in some godforsaken village in Portugal, be it the appraisal of a given railway, be it the improvement of the ministry's latrines. So the Junta had neither the time nor the resources to undertake the duty of devising a national railway

[3] Ministry of Public Works Historical Archive (henceforth MPWHA), Junta Consultiva de Obras Públicas e Minas (henceforth JCOPM), book 32 (1871), 280-299; book 32-A (1871), 1-8v; book 34 (1873), 97v-100v; book 35 (1874), 149. Associação dos Engenheiros Civis Portugueses, 1871-1872.

plan. They only went so far as to suggest a couple of lines amidst a report about two other railways.[4] This, notwithstanding, was very revealing of the importance the engineers gave to the general network plan. Seldom did they in their reports go further than what the government had asked them. If the minister asked about a road, they would only give their advice about that road, even if other details could be justifiably included in the report.

In any case, if the mission of studying and proposing a railway network was too heavy for the Junta, it could be handled by another institution: the Civil Engineers Association. The debate began in 1876, when the Association board nominated a committee (Lecocq, Afonso de Espregueira, Xavier Cohen, Inácio Lopes and Frederico Pimentel) to study and propose a draft of the desired general network plan. By then Portugal had about 900 kilometres of railways spread across the country (Pimentel 1892; Valério 2001, 372). It was unanimously accepted that the network was flawed as far as routes and traction conditions were concerned and it had been devised to serve mainly political purposes. Therefore, any proposal could simply try to mitigate those errors and avoid them in the construction of new railway lines (Larcher 1878, 269). The first plan formulated by the aforementioned engineers was presented to their colleagues in May, 1876, merely three months after their assignment, and it opened the floor for discussion (the proceedings of some of these reunions were published in the *Revista*, but the majority was lost). In the following months, several other technicians expressed their thoughts about this issue, suggesting changes to the original plan and the committee itself was rearranged during the process (some engineers were added to the original list of committee members). In February 1877, all the proposals were taken into consideration by the committee that altered its original proposal. This was followed by yet another five-month period of debate that ended with the publication of the committee's final draft of the general network plan, which was in all likelihood delivered to the government. This, however, did not put an end to the discussion, for afterwards some engineers were still arguing about the defects of the committee's suggestions. Men like Frederico Augusto Pimentel, Jaime Larcher or Miguel Carlos Correia Pais were not happy with the final result of the debate and they kept expressing their opinion both in the Association and the *Revista de Obras Públicas*. Even some army officers added a few thoughts to this dispute, even though they considered the issue mainly from the point of view of the defence of the country, a detail that had been kept aside in the Association's work. Despite the fact that

[4] MPWHA, JCOPM, box 18 (1875), report 6418 (January 7).

most engineers had a military background (the majority graduated from the Army School) (Macedo 2009), it was decided that since the Association was of civil engineers, any sort of military reasoning should not be taken into consideration (as a matter fact, the vast majority of the army's officers advices were not taken into consideration by the government) (Pereira 2012b).

Back to the Engineers Association, the debate had lasted for a year and a half and it ended with the publication of the committee's final report. In this 18-month discussion the engineers could participate in either two ways: those who were in Lisbon could attend the meetings taking place in the Association's headquarters; those who were stationed far from the capital and could not attend those reunions could send their reports by mail. Naturally their influence over the final decision was weaker, but they were very important due to their knowledge of the districts to which they were assigned.

Anyhow, in the late summer of 1877, the final official report of the Association was ready. The network should be divided into three classes of importance. The report privileged international connections. Two of the first class railways were supposed to cross the border with Spain: the whole Portuguese railway strategy was based upon the goal to take advantage of Portugal's position in the Western most corner of Europe and turn its harbours, namely Lisbon's, in the main gateways between Europe and the rest of the world (Alegria 1990; Gomes 1998; Pinheiro 1986). Nevertheless, there was also a set of internal lines that were expected to expand the benefits of railways to the whole country and to increase the traffic in the existing railroads. Those lines filled the second and third classes. The difference between these two ranks was the gauge: second class railways should be built in broad gauge (1,667mm or 5' 5 ¾'') while third class railways should be narrow gauged. The exact value of the narrow gauge was yet to be decided, but it should be applied uniformly throughout the entire third class network.

This report should avoid the errors committed during the planning and construction of the existing lines, namely those with political basis. However, one can be suspicious about the motivations behind the inclusion of some lines in the Association's proposal. For starters, the Cáceres embranchment between the Spanish province with the same name, the border between Portugal and Spain and the Eastern line between Lisbon and Elvas. This railway only interested the Royal Company of Portuguese Railways, a private company with French capital and, as it happens, two of the members of the committee (Afonso de Espregueira and Veríssimo Guerreiro) were employees of the Royal Company. Even though the Cáceres embranchment was considered useful by a few of the engineers, it was never considered a first class railway. For the purpose it served (a quicker connection between Lisbon and Madrid), another line a bit further to the North—through the province of Beira Baixa—was preferred). From a military point of view, it was even regarded as a disaster, since it allowed a connection between Portugal and Spain avoiding the fort and the artillery of Elvas. The Royal Company was also interested in a shorter route to Madrid, but was not able to build the Beira Baixa railway (much more expensive and harder to set) and, on the other hand, it hoped to take advantage of the mining industry of that Spanish province (Pereira 2011).

Another example was the railroad through the Minho region between Braga and Monção, suggested by the very same Veríssimo Guerreiro. It was a line that no engineer but Guerreiro had ever advised. Well, if we analyse his biography we realise that Veríssimo Guerreiro was in fact from that region and he probably wanted to benefit the area where he had been brought up.[5]

Back to the debate, it ended with the elaboration of a final report that was delivered to the government. However, all effort was in vain, since no general plan would be approved in the following years and not until the eve of the twentieth century. Furthermore, and despite the debate had lasted for almost two years, the minister of Public Works (Lourenço de Carvalho, who had also taken part in that discussion) decides to ask for yet another opinion and he charges his colleague João Crisóstomo de Abreu e Sousa with the task of drawing a new network. Lourenço de Carvalho could not agree with the draft of the Association since it did not propose a line through the Douro Valley (where Lourenço's family had quite a few expanses of land). João Crisóstomo accepts the task and in 1878 he presents yet another survey to the minister. Finally, in 1879, Lourenço de Carvalho proposes his project to the parliament. It was pretty similar to Crisóstomo's essential ideas, although one can perceive a few differences between the two plans (Alegria 1990). In any case, Lourenço's plans would never be approved in any of the two chambers of the Portuguese parliament. According to parliament regulation, the bill should be sent to specific commissions (in this case, the Public Works, Finance and eventually War commissions), that in turn would write down a report to be debated in both parliamentary houses. Nevertheless, the commissions never wrote any statement and the bill was lost in the archives of the parliament (Fino 1883-1903, vol. 1, 375-393; Santos 1884; Santos 1986).

---

[5] MPWHA, individual files.

The approval of a general network plan was unappealing from a political standpoint, because it would bind the action of any government and the cabinet would have to keep to the lines mentioned in that plan. If the opportunity to build any other railway would arise, the government would be forced to ask for an alteration of the law in the parliament, submit itself to the criticism of the opposition and maybe jeopardise the success of the hypothetical negotiations with the private investor(s). It could also prevent the government from making deals with their political partisans. For these reasons, a network plan was never approved nor discussed in the parliament until 1898, even though many congressmen had demanded it. However, when these men became ministers they conveniently forgot those demands and kept the plan out of their parliamentary agenda.

Only in 1898 would the parliament issue a law defining the general plan. In that year Portugal had already about 2,300 kilometres of railways and so the plan could not perform a very comprehensive job. In fact, it could only determine the construction of second class railways, since the basis of the national rail network was already set. The July 14, 1898 law instructed three commissions to study this issue and propose three regional plans: one for the North of Portugal; the second for the South; the last for the region between the Mondego and Tagus rivers (they would be issued and approved by government decrees between 1900 and 1905). In order to perform their task, the commissions were required to ask for the input of the district councils, class and business associations and, of course, the Portuguese Civil Engineering Association, whose advice was the most considered by the commissions, which is not at all surprising once one recalls that all of the commissioners were members of the Association. In any case, as it was already mentioned, this plan could never be of as much importance as of the ones debated in the 1870s, since the main part of the network was already set by the end of the nineteenth century. On the other hand, we cannot ascertain how comprehensive was the debate in the Association or if it even took place, since there are no records of the process that led to the final report issued by the Association.

The Association's contribution to solutions in hilly regions

Building railways in the Iberian Peninsula could be a problem due to the ruggedness of the territory. As a matter of fact, in the 1840s several Spaniards doubted that it could even be possible to build railways in their country (Moreno Fernández 1996; Pinheiro 1986; Santos 2011). In Portugal this problem was greater in the Northern provinces. In the South, theoretically, it would be easier to set rails, because the land was flatter. It is no surprise that the construction began precisely there (in the 1850s the government signed three contracts to build three railways in the Southern regions of the kingdom).

But as years went by and as the construction was being taken up to the North, the need to overcome its narrow valleys and rugged surface became a concern for the government and the engineers. The first major lines built in the North (the construction of the Minho and Douro lines began in 1872) were, nonetheless, broad gauged. It was believed that they would be part of the first class network and they should have the same gauge as the other railways already open for service. Besides, since those lines were supposed to be the natural continuation of the Northern line, it made no sense to give them a different distance between rails. However, it was expected that, during the construction, the engineers hired by the State (who was the contractor of these lines) could drop the costs of construction. As for the remainder of the Northern network, broad gauge was not compulsory—far from it, actually. The cost of building railways in the hills of Trás-os-Montes (the most inland province of the North), for instance, could be very high, or so it was believed. Therefore, another solution had to be found and that solution was narrow gauge.

This sort of railroad was the main interest of the Portuguese engineer Cândido Celestino Xavier Cordeiro during his stay in Paris where he attended the École de Ponts et Chaussés in the late 1860s (Matos 2009). In the early 1870s he is charged by the government with the task of studying this kind of railways in Central European countries (Germany, Austria and France). In 1871 and 1872 a rather comprehensive study about the advantages of narrow gauge (probably written by another engineer, Vitorino Dámasio, according to his biographer) is published in the *Revista de Obras Públicas*.[6] At the same time, the Association began advertising the usefulness of narrow gauge railways abroad through that same journal. By then, the government and the Portuguese engineers had realised (with the example of the Minho and Douro lines) that building broad gauge railways in the North would be tremendously expensive and narrow gauge could be a valuable solution. Outside the Association, in the ministry, the engineers on its advisory board were also in tune with their colleagues. They believed that narrow gauge could be a viable option in the hilliest regions of Portugal where high traffic was not expected. As we mentioned before, they only advised that the gauge should be identical throughout the

[6] Associação dos Engenheiros Civis Portugueses 1871-1872

country and that a general network plan should be approved before granting any new concessions.

This brainstorming encouraged private investors to present their projects to the government (according to the December 31, 1864 decree, no railway could be built without the authorisation of the government and in some cases of parliament). Until 1875, nine proposals were submitted to the Ministry of Public Works, most of them outside the hilliest regions of Portugal. However, none was carried out. The government did not offer any sort of financial aid and the entrepreneurs were not able to gather the necessary capital to accomplish their goals. Besides, it is also likely that these men wanted no more than to get a hold on the concession and then sell it to the highest bidder. At the time Portugal knew a boom in banking activity and speculation reached an all-time high (Pereira 2012a; Valério 2006).

In any case, the Ministry of Public Works realised that public powers were obliged to aid these entrepreneurs, otherwise their proposals would never be more than a few lines in a piece of paper (mainly after the 1876 crisis brought some banks to their knees). Furthermore, the example of the only narrow gauge railway in Portugal was not appealing. It was the line between Porto and Póvoa de Varzim (in the coast, a rather flat area, so this railway did not really serve as an example of how narrow gauge could overcome the obstacles in rugged terrain), open for business since 1875. Even though the cost of its construction had been rather low, its profits were not high enough to attract new investments (Pinheiro 1986; Vieira 1983).

The first decision taken by the government was to promote surveys. Therefore, in the following years, the debate was resumed and new studies for the Northernmost Portuguese provinces (Minho and Trás-os-Montes) were ordered. In 1878, commissioned by the government, Francisco Maria Sousa Brandão presents a proposal for a narrow gauge network in Trás-os-Montes, after spending some weeks in Italy and Switzerland. A few months later, Pereira Dias publishes an essay about a network in the neighbouring region, Minho. He had been charged by some municipalities of that province (Brandão 1880; Dias 1881). Both these studies were endorsed by the Association and published in the *Revista de Obras Públicas*, which kept printing news about how well narrow gauge railways were doing abroad. Xavier Cordeiro was once again sent abroad to expand his knowledge on this matter. Afterwards, he published a small book about narrow gauge, which became the most comprehensive essay about this sort of railroad. Once more, the Association completely endorsed this study and published it in several issues of its *Revista* (Cordeiro 1879).

Despite all this advertising, nothing was actually being done in the field. It is a fact that a handful of projects were proposed to the government, but the cabinets still refused to provide any kind of financial assistance to these entrepreneurs. The company that operated the line between Porto, Póvoa de Varzim and Famalicão also had an idea to extend its railways towards the heart of Trás-os-Montes. However, it demanded from the government a guarantee of income, which was refused.

Narrow gauge railways were supposed to be low-cost and several ministers and engineers hoped that those costs could be supported entirely by private investors. Hence, the exchequer could focus on aiding broad gauge railways, in theory more expensive and unattainable. However, it was soon realised that without financial support, no narrow gauge would ever be built in the hilliest regions of Portugal.

In the early 1880s the first narrow gauge resulting from a joint venture between private investors and the State would emerge. It was a railway built in Portuguese India (state of Goa), that was to connect the harbour in Mormugão with the British India rail network over the Indian mountain range known as the Western Ghats. The parliament decided to approve the government's resolution to support with a guarantee of interest the efforts of the Stafford House Committee (that would later fund the West of India Portuguese Guaranteed Railway Company). This event would become an important landmark in the history of narrow gauge railways in Portugal. The Goan territory was almost unknown to the Portuguese authorities and to the British engineers of the company. An urgent survey of the region, namely the mountains of the Western Ghats, was in order and the Portuguese government decided to send its expert in narrow gauge—Xavier Cordeiro—to work alongside the British engineers. Afterwards, he became fiscal director of public works in India and was assigned to survey the railway construction. This was the first time Portugal built a narrow gauge railway in hilly regions. The construction was not easy and it took longer than expected. Meanwhile, the Portuguese minister of Marine chastises the work of Xavier Cordeiro as fiscal director. Cordeiro resents the allegations and resigns his commission in India and as an engineer in the Ministry of Public Works, later being hired by a Portuguese private railway company—the Royal Company (Kerr and Pereira 2012).[7] Nonetheless, he did not quit his membership in the Engineers Association. After all, he still was an engineer working in a railway company. Therefore, it is likely that he shared his experience in India with his fellow colleagues of the Association.

[7] MPWHA, individual files

At the time (early 1880s), Portugal was also building another three narrow gauge tracks, this time on the mainland: in the province of Trás-os-Montes up the Tua River; in Minho around Guimarães; and in Beira Alta around Viseu. The Tua railway was one of the lines mentioned in the report Sousa Brandão delivered to the government in 1878 and it was supposed be the first step to endow the Northeast of Portugal with railways. However, until the end of the century no more narrow gauge tracks would be set in the kingdom, despite all the efforts made by the governments and supported by the Association (Davies 1998).

In 1888, the government presented a bill to the parliament to build several narrow gauge railways in the North of Portugal.[8] This bill was supposed to solve once and for all the transport problems in that part of Portugal. Hundreds of kilometres of track would be laid. In the *Revista de Obras Públicas*, the engineer Augusto Miranda Montenegro showed his support to the government proposal and it is plausible that other engineers in the Association shared his opinion (Montenegro 1889). We can find evidence for this if we take a look at another periodical issued in this year, the *Gazeta dos Caminhos de Ferro de Portugal e Hespanha* (*Railway Gazette*, for short). Several engineers that worked on railways believed that the government's proposal was indeed very good and they were very disappointed when the bill did not come through. As a matter of fact, it was not even sent to the parliamentary commissions, which was the procedure that began the parliamentary debate process. Previously the opposition had made quite clear that it did not believe that the scheme was even feasible in financial terms. Furthermore, it accused the government of electioneering, stating that the purpose of the bill was merely to conquer more votes in the upcoming elections (Pereira 2012a, 141-142).

It is a fact that Portugal was in no shape to uphold such ambitious investment (four years later the exchequer would declare a partial default), but the whole situation was seen a few years later as a lost opportunity. Actually, that was in part the spirit of the aforementioned law of July 14, 1898. The purpose of the 1888 bill was legitimate, but the means to get it were not (back then the bill suggested granting several guarantees of income to private companies; in 1898 those guarantees were replaced by a financial scheme involving several public revenues). What did not change was the trust in narrow gauge. In fact, one of the main goals of the law was to build railways in the hilly and poorer regions of Portugal, where narrow gauge was the wisest choice. As we mentioned, the Engineers Association was also largely favourable to this project, despite the fact that, by that

[8] Diário da Câmara dos Deputados, June 1, 1888, 1813-1819.

time, in countries like the United States the supposed advantages of narrow gauge were being thoroughly questioned (Puffert 2000). After the law was issued, the government nominated three commissions that should draw a draft of the network and then ask the opinion of local powers and other associations, including the Engineers Association. The final result was a railway map where narrow gauge would play a very important role. In the following years and until the 1940s several of those lines were actually built, but others were nothing more than a hair stroke on a paper (Pinheiro, Lima and Paulino 2011; Santos 2011).

Conclusion

The construction of railways in Portugal began in the 1850s, long before the Portuguese Association of Civil Engineers was created. But in the late 1860s, the public works programme was halted and the engineers decided to gather in a society to lobby for the continuation of railway construction.

From that moment on, the Association became an important partner of the several cabinets that ruled the country. Its goal was quite clear: to promote the dissemination of practical knowledge about engineering. In the period between 1870 and the end of the century, the Association did precisely that. As far as railways were concerned, its members promoted comprehensive debates about several railway issues. Their intellectual production was then published in the Association's own journal, the *Revista de Obras Públicas e Minas*.

However, if the engineers were unanimous in believing that Portugal needed more and more railways, there was no consensus when the debate narrowed down to which lines should be built and what their features should be. Some of these disagreements were based upon different expectations about railways, but they were also based upon political and business reasoning. Some engineers were also politicians, either congressmen or ministers; others were employees of influential railway companies; but all had been born and raised in different areas of Portugal and they would feel tempted to promote railways in those regions instead of others. Therefore, when the level of discussion came to details, the consensus was gone.

This was beneficial to the government, which was able to pursue its own agenda without going against the judgement of the Engineers Association, for it was very easy to find at least one engineer that could back the decisions of the cabinet. In Portugal, political power overcame technical authority, which was made possible due to the lack of harmony amongst the engineers. Nevertheless, the Portuguese Civil Engineers Association played a very important role in the debate about railways in Portugal. Through its magazine—the *Revista de Obras Públicas e Minas*—it divulged in Portugal the development of railways abroad and contributed with several essays, surveys, studies and solutions for railway building in the nation.

Sources

Associação dos Engenheiros Civis Portugueses. 1871-1872. "Caminhos de ferro economicos." *Revista de Obras Públicas e Minas* 21-25: 315-447 and 1-22.

Brandão, Francisco M. de S. 1880. "Estudos de caminhos de ferro de via reduzida ao Norte do Douro." *Revista de Obras Públicas e Minas* 125-126: 145-183.

Collecção Official de Legislação Portuguesa (1845-1892). Lisbon: Imprensa Nacional.

Cordeiro, Cândido C. X. 1879. "Memoria ácerca dos caminhos de ferro de via reduzida". *Revista de Obras Públicas e Minas* 113-115: 237-365.

Diario da Camara dos Deputados. Lisbon: Imprensa Nacional.

Dias, João J. P. 1881. *Memoria ácerca dos caminhos de ferro de segunda ordem no districto de Braga*. Lisbon: Imprensa Nacional.

Dinis, Pedro G. S., compil. 1915-1919. *Compilação de diversos documentos relativos à Companhia dos Caminhos de Ferro Portuguezes*. Lisbon: Imprensa Nacional.

Fino, Gaspar C. G. C., compil. 1883-1903. *Legislação e disposições regulamentares sobre caminhos de ferro*. Lisbon: Imprensa Nacional.

Larcher, Jaime. 1878. "Caminhos de ferro em Portugal." *Revista de Obras Públicas e Minas* 102-103: 256-279.

Ministry of Public Works Historical Archive. Individual files.

—. Junta Consultiva de Obras Públicas.

Montenegro, Augusto P. M. 1889. "A rede complementar dos caminhos de ferro ao norte do Mondego." *Revista de Obras Públicas e Minas* 237-238: 315-341.

Revista de Obras Públicas e Minas. Lisbon: Imprensa Nacional.

Santos, Clemente J., compil. 1884. *Caminhos de ferro. Pareceres parlamentares de 1845 a 1884*. Lisbon: Imprensa Nacional.

Bibliography

Alegria, M. F. 1990, "A organização dos transportes em Portugal (1850-1910): as vias e o tráfego." PhD diss., University of Lisbon.

Bonifácio, M. F. 1992, "A guerra de todos contra todos (ensaio sobre a instabilidade política antes da Regeneração)". *Análise Social* 115: 91-134.

Branco, R. 2003, *O Mapa de Portugal. Estado, Território e Poder no Portugal de Oitocentos*. Lisbon: Horizonte.

Brito, J. M. B., M. Heitor and M F. Rollo, eds. 2003, *Engenho e Obra. Uma abordagem à história da Engenharia em Portugal no século XX*. Lisbon: D. Quixote.

Davies, W. J. K. 1998, *Narrow Gauge Railways of Portugal*. (S. l.): Plateway.

Dias, M. H., ed. 1995, *Os Mapas em Portugal. Da tradição aos novos rumos da Cartografia*. Lisbon: Cosmos.

Diogo, M. P. 1994. "A construção de uma identidade profissional. A Associação dos Engenheiros Civis Portuguezes (1869-1937)." PhD Diss. Nova University of Lisbon.

—. 2003, "Engenharia e Modernidade. Os Engenheiros Portugueses entre as Obras Publicas e a Indústria." *Boletim da Sociedade Portuguesa de Química* 88: 13-17.

Gomes, G. 1998, *A Luz e as Sombras na Gare do Oriente. O Caminho de Ferro no final do séc. XX*. Lisbon: CP.

Guillemois, I. 1995, "Les transports au Portugal au XIX siècle (de 1843 à 1899) à travers les Viagens na Minha Terra de Almeida Garrett et As Farpas de Ramalho Ortigão." Masters diss. University Michel de Montaigne, Bordeaux.

Justino, D. 1988-1989, *A formação do espaço económico nacional. Portugal 1810-1913*. Lisbon: Vega.

Kerr, I. & H. S. Pereira, 2012, "Late nineteenth century exploitations: comparing Mormugão (India) and Tua railways." In *Railroads in Historical Context: construction costs and consequences*, ed. A. McCants, E. Beira, J. M. L. Cordeiro and P. B. Lourenço, forthcoming. Porto: MIT, EDP, University of Minho.

Macedo, M. C. 2009, "Projectar e construir a Nação: engenheiros e território em Portugal (1837-1893)." PhD diss. University of Coimbra.

Mata, M. E. & N. Valério, 1993, *História económica de Portugal. Uma perspectiva global*. Lisbon: Presença.

Matos, A. C., & M. P. Diogo, 2006, "From the École de Ponts et Chaussées to Portuguese railways: the transfer of thecnological knowledge and practices." Paper presented at the Second International Conference of the Association Internationale pour l'histoire des Chemins de Fer. Railway modernisation: an historical perspective, Lisbon, November 27-29. Accessed February 18, 2011. www.docutren.com/archivos/lisboa/pdf/22.pdf

Matos, A. C., & M. P. Diogo, 2007, "Bringing it all back home: Portuguese engineers and their travels of learning (1850-1900)." *Journal of History of Science and Technology* 1. Accessed March 7, 2011. www.johost.eu.

Matos, A. C. 2009, "Asserting the Portuguese Civil Engineering Identity: the Role Played by the École des Ponts et Chausées." In *Jogos de Identidade Profissional: os Engenheiros entre a Formação e a Acção*, ed. A. C. Matos, M. P. Diogo, I. Gouzevitch and A. Grelon, 177-208. Lisbon: Colibri.

Matos, A. T. 1980, "Transportes e comunicações em Portugal, Açores e Madeira: 1750-1850." PhD diss. University of Azores.

Moreno Fernández, J. 1996, *El ancho de vía en los ferrocarriles españoles: de Espartero a Alfonso XIII*. Madrid: Editorial Toral.

Pereira, H. S. 2011, "Caminhos-de-ferro da Beira (1845-1893)". *Revista de História da Sociedade e da Cultura* 6: 273-296.

—. 2012ª, "A política ferroviária nacional (1845-1899)." PhD diss. University of Porto.

—. 2012b, "Portugal, Espanha e as preocupações militares sobre caminhos-de-ferro (1845-1899)". Paper presented at the Sixth Railway History Conference of the Fundación de los Ferrocarriles Españoles, Vitória-Gasteiz, September 5-9. Accessed October 8, 2012. www.docutren.com/congreso_vitoria/comunicaciones/4036.pdf

Pimentel, F. A. 1892, *Apontamentos para a historia dos caminhos de ferro portuguezes*. Lisbon: Tipografia Universal.

Pinheiro, M. 1986. "Chemins de fer, structure financiere de l' État et dependance éxterieure au Portugal: 1850-1890." PhD diss. University of Paris.

—. 2008. *Cidade e caminhos de ferro*. Lisbon: Centro de Estudos de História Contemporânea Portuguesa.

Pinheiro, M., N. M. Lima & J. Paulino, 2011, "Espaço, tempo e preço dos transportes: a utilização da rede ferroviária em finais do século XIX." *Ler História* 61: 39-64.

Puffert, D. 2000, "The Standardization of Track Gauge on North American Railways, 1830-1930." *The Journal of Economic History* 60-4: 933-960.

Rodrigues, M. L. 1999, *Os Engenheiros em Portugal*. Oeiras: Celta.

Santos, L. A. L. 2011, "Politica ferroviaria ibérica: de principios del siglo XX a la agrupacion de los ferrocarriles (1901-1951)." PhD diss. Complutense University of Madrid.

Santos, M. P. 1986, *Monarquia Constitucional. Organização e Relações do Poder Governamental com a Câmara dos Deputados (1834-1910).* Lisbon: Parliament.

Valério, N. ed. 2001, *Estatísticas Históricas Portuguesas*. Lisbon: INE.

Valério, N. ed. 2006, *História do Sistema Bancário Português*. Lisbon: Bank of Portugal.

Vieira, A. L. 1983, "The role of Britain and France in the finance of Portuguese railways 1850-1890. A comparative study in speculation, corruption and inefficiency." PhD diss. University of Leicester.

CHAPTER FOUR

THE GEOGRAPHICAL SOCIETY OF LISBON AND THE NEW CHALLENGES IN THE SCIENCE OF THE NINETEENTH AND TWENTIETH CENTURIES

RUI PINTO

Introduction

The Lisbon Geographical Society (SGL) was founded in 1875 following a movement that spread across Europe, creating institutions of a similar nature, namely in Paris, Berlin and London. Luciano Cordeiro and an ample number of intellectuals of that period were responsible for the foundation of the Lisbon Geographical Society.

The Portuguese government placed the "Permanent Central Commission for Geography", that had been created in the Ministry of the Navy and Overseas Possessions, by decree on February 17, 1876, within the Society, with the objective of "organising scientific expeditions, collecting samples and documents that were relevant to the development and improvement of geography, ethnological history, archaeology, anthropology and natural sciences in relation to the Portuguese territory, namely the overseas possessions, promote and cooperate with any projects related to these sciences, and propose to the government all measures to ensure more and better knowledge of those vast and important regions".

Thus the Society was born for the "study, discussion, teaching, investigation, and scientific expeditions of the various branches of geography, principles, relations, discoveries, progress and applications".

Up to this day the Lisbon Geographical Society has been a reference in scientific investigation in practically all branches of knowledge. We shall analyse their most relevant contributions in the last 136 years.

The early days

From the beginning of the second half of the nineteenth century, European rivalries led to an unprecedented race to the interior of Africa. Urgent completion of cartographic information was required due to the new market demand for raw materials that could be claimed by those that had "historical rights" based on sovereignty.

Between 1802 and 1811, Silva Porto, with the help of the slave traders Pedro João Baptista and Amaro José, made the third trip to the eastern African coast. Here he contacts David Livingstone for the first time, who questions his scientific knowledge.

Silva Porto gave his answer in 1868, in a document found in his belongings, published after his death by the Geographical Society of Lisbon in 1891:

> The Reverend Dr. David Livingstone earned, without a doubt, the crown that his fellow citizens gave him for services rendered in these parts of Africa; however, truth be told, it was shattered by the illustrious traveler when he tainted it with the filth of slander.
>
> He desired to reach his goals with scant regard for the means.
>
> In April or March 1853, on the day that he heard of me, a bolt of lightning striking close to him would not have caused the same impression that such news did, but surely he would understand that, sooner or later, he would face a competitor, obscure because of his poor talent, yes, but a living witness of the priority on the same places that the doctor considered his right to call himself the first European visitor. It does not belong to me entirely, it is true, seeing as that other people have gone through these places before me and long before the illustrious traveler, but in fact it belongs to me, because those people were sent by me, they existed, and still exist, presently in greater number in my service, some from Luanda, others from the High Golungo, others from Ambaca, others from Pungo-Andongo, others, finally, from Bihé.
>
> Has not the illustrious author of the Exam proved exhaustively that it belongs to the Portuguese since ancient times?(Porto, 1891)

Between 1852 and 1860, the Austrian Frederick Welwitsch, paid by the Portuguese government, proceeds with botanical studies in Angola (Graça, 1867; Boletim da Sociedade de Geografia de Lisboa, 1890).

With the support of the railroad and under the direction of Andrade Corvo, Minister for Overseas, plans for the development of the African corridor are thoroughly studied and prepared, prompting a series of expeditions that would decidedly mark the future political position on the African continent.

The Englishman Verney Lovett Cameron is sent by the Royal Geographical Society in 1873 to assist Livingstone. With his death, Cameron becomes the first European to cross Africa from coast to coast.

The Belgians enter the race when Leopold II, facing rapid demographic growth in his country, decides to dispute the African markets, to the point where, without inviting Portugal, he called for an International Conference on Geography in 1876, that gave rise to the International Association of Africa; supposedly a scientific and humanitarian organisation that aimed to promote the development and civilisation of Central Africa, but that in reality had political aims. He engages with the American journalist Stanley in order to call attention to the International Association of the Congo.

In Portugal, interest in the African continent had for a long time taken a back seat in favour of Brazil. The efforts of the Marquis Sá da Bandeira allowed the defense of Portuguese interests in Africa to be put back on the political agenda, through administrative and economic measures, namely the abolition of slavery, which met with strong opposition from the African slave traders.

Luciano Cordeiro ponders the creation of a scientific association, similar to others that appeared in most parts of the world, like the one in Paris in 1821, Berlin in 1828, London in 1830, St Petersburg in 1845, Vienna and Rome in 1856 and Bucharest in 1875. During this period the main purpose of these associations was to organise expeditions and present their results through conferences, exhibitions or periodicals. In most cases this was a race for knowledge, in order to justify political ambitions of governments that supplied the necessary support (financial, technical, and sometimes military).

The teacher at the Lisbon Industrial and Commercial Institute, Rodrigo Afonso Pequito, the architect Emiliano Augusto de Bettencourt,the engineer João Cândido de Morais and the philosopher Dr Cândido de Figueiredo became the inner core for the creation of a Geographical Society.

The idea quickly spread through the Lisbon literary circles, as both Luciano Cordeiro and Rodrigo Afonso Pepito were frequent visitors at the Livraria do Carmo bookshop and the newsrooms of the *Diario de Notícias* and the *Paiz*. They were looking for older people that could add credibility to the project: 'It could not be exclusively the work of the young' (Pequito, 1901).

They then propose to the king D. Luís I the creation of a "Scientific Society (…) dedicated to promoting and aiding the study and progress of geographical and related sciences, in the country." (*Boletim da Sociedade de Geografia de Lisboa*, 1950). Rodrigo Afonso Pequito accompanied him in delivering the petition that listed various personalities: 63 ordinary founding members and 11 correspondent founding members.

The Society aimed also to embrace positivism and extend the institution abroad in what was described as "scientific conviviality of the civilized world", in a subliminal message to the masonic values "this great light that upon us all shines and cheers (…) the light and the justice of modern Critic" highlighting the figures of the maritime discoveries of "cosmographers and navigators" (Sociedade de Geografia de Lisboa, 1950).

What was at stake was the development and the applicability of the new investigations and scientific explorations not only from a geographical point of view but also historical and ethnographical. The means for reaching the aims of the Society would be:

- First: Sessions, conferences, speeches, open courses, competitions and scientific congresses, scholarships for studies and investigation, trips, etc.;
- Second: Publications, creation of libraries, archives and museums, correspondence and relations with the various scientific centres and guilds.

D. Luís and D. Carlos declared themselves protectors of the society by Royal Letters on November 14, 1878 and September 27, 1890. In the letter dated November 14, D. Luís refers to the "laudable determination given to the indispensable efforts to develop geographical studies and explorations".[1] Plans for the progressive occupation of Africa began to worry the Portuguese government regarding their domains.

The Society's statutes were created on November 10, 1875, subscribed by 74 men; five of whom were part of the provisional committee. However, less than a month later, the Ministry of the Navy and Overseas created, by decree, another body called Commissão Central Permanente de Geographia (Permanent Central Commission on Geography).

The arrival of this Commission led to a malaise among members of the Society, since their purpose was practically identical. In 1880, the former president of the Geographical Society of Lisbon, Viscount of São Januário, becomes the new Minister of the Navy and Overseas and, under his influence, the Commissão Central Permanente de Geographia is placed within the Society and is awarded an advisory role to the government.

---

[1] Available at http://www.socgeografialisboa.pt/historia

After the foundation of the National Committee for the Exploration and Civilisation of Africa, the Geographical Society promoted and sponsored several expeditions to explore and occupy the territory between Angola and Mozambique: by Hermenegildo de Brito Capelo and Roberto Ivens, from 1877 to 1880 and from 1884 to 1885, to the North of Mozambique; by Major Serpa Pinto and naval ensign Augusto Cardoso, from 1877 to 1879; by Henrique Augusto Dias de Carvalho to the East of Angola, from 1884 to 1888; and by the naturalists Francisco Newton, from 1880 to 1907, and José Pereira do Nascimento, from 1888 to 1911, to the West coast of Africa.

Ernesto de Vasconcelos, president of the Cartography Commission and Perpetual Secretary of the Lisbon Geographical Society between 1911 and 1930, encouraged the founding of various scientific and geographical missions, as well as the drawing up of various charts of the overseas territories.

He published a substantial amount of documents on the subject of border delimitation that argued in favour of the possession of those territories. He was also the organiser of the Cartography Exhibition (1903-1904), the Exhibition of Colonial Goods (1903 and 1906), the Exhibition of Means of Transport (1909) and the Colonial Congresses of 1901, 1924 and 1930. In 1904 the Geographical Society of Lisbon organised the exhibition of the oceanographic collections of King D. Carlos I.

The statutes of 1876, still in force, initially set up the existence of Permanent General Commissions (African, American, Asian and Islands of Emigration), as well as Professional Sections (Mathematics, Physics, Chemistry, Astronomy, Meteorology, Geology, Geodesy, Mineralogy, Botany, Zoology, Anthropology, Ethnography, Agriculture, Industry, Commerce, Political Economics, Statistics, Social Sciences and Jurisprudence, Art and Literature, History, Engineering and Land Communications, Military Science, Naval Science and Maritime Communications, Medical Sciences, Cartography, Study of Geography, and National Public Education).

In 1878 authorisation was given to the Project of a School of Subjects Relating to the Land and the People and the Languages of Overseas Portugal–Portuguese Colonial Course, to prepare those who were sent for public service in the Colonies.

But it was only in 1906 that a Colonial School was formally created, under the Ministry of the Navy and Overseas, and moved to the Geographical Society of Lisbon. Later that was the origin of the present day Higher Institute of Social and Political Sciences of the University of Lisbon.

In a similar way the Higher School of Physical Education was created, part of the SGL between 1930 and 1940, before turning into the National Institute of Human Motricity, also assimilated by the same university.

In 1927, the Geographical Society of Lisbon holds the first of the Colony Weeks, in collaboration with the General Colony Agency, universities, secondary schools, local councils and various associations, such as the "Pro-Colony Movement" from Lisbon and Oporto. Three big events for publicising the colonies also mark this period: the Industrial Exhibition of Lisbon of 1932; the Portuguese Week in Vigo; and the Estoril Casino Conference of 1933. In 1934 the Great Colonial Exhibition of Oporto is held and the Society participated by organising the Colonial Congress.

In a circular dated February 15, 1884, the Geographical Society of Lisbon informs its members that the process of creating a Colonial Museum had begun. However, it was only in 1892 that the government placed the Naval Colonial Museum (created in 1871 and belonging to the Ministry of Naval and Overseas Affairs) in the Geographical Society, hence known as the Colonial and Ethnographic Museum.

After the voyages of exploration to Angola and Mozambique undertaken by Hermenegildo de Brito Capelo and Roberto Ivens, from 1877 to 1880, the Lisbon Geographical Society publishes a map where much of Central Africa now appears under Portuguese rule and also sets a public subscription for the establishment of civilising stations in this entire African region. However, this project would not receive any public support from the Portuguese government.

From the Treaty of Zaire to the Ultimatum

Under the Treaty of Zaire, of February 26, 1884, Britain recognised the Portuguese hegemony on the mouth of the river Zaire, but taking as counterparts full freedom of trade and navigation on the rivers Zaire, Zambezi and their tributaries and tax-free shopping in the realm of Congo, leaving Portugal with no chance of competing commercially. But with the opposition of the French and German governments, and King Leopold of Belgium, on account of the interests they had in this territory, this agreement was blocked. The British supported the abolitionist cause in Africa in order to penetrate more easily in this market.

The European powers England, France, Germany, Holland and Spain and the United States itself were unhappy with the results and these countries strongly protested, demanding the annulment of the treaty because of: "the ancient historical rights of Portugal being negligible in the area" (A.V. 1989).

England as such is forced to renounce it. The Portuguese Foreign Minister, José Vicente Barbosa du Bocage (President of the Geographical Society of Lisbon from 1877 to 1883), suggested to the English Government to organise an international conference aimed at solving the problem.

In March 1887, Captain Alfredo Augusto Caldas Xavier volunteered to undertake a new exploration, in Africa, which was supported by the African Commission of the Geographical Society of Lisbon.

With the rise to power of the progressive government of José Luciano de Castro, the pro-German Henrique Barros Gomes, first as minister of the Navy and Overseas and afterwards as Minister of Foreign Affairs of the progressive government, submitted to the House of Representatives in 1887 a map of southern Africa, known as the pink map (Cor-de-Rosa) and dated 1886. It backed the intentions of the Geographical Society of Lisbon, of which de Castro also was a member, to promote the expeditions that could ensure this project.

The Geographical Society of Lisbon believed that the best way to enforce Portuguese rights in Africa would be the construction of railways (see Chapter Three of this book) and the extension of telegraph lines which would allow the commercial development of the region, namely mining. So, in November 1889, a proposal to create a network of telegraph lines between Quelimane, Niassa, Tete, Zuiribo, Manica, Beira, Sofala, Inhambane and Lourenço Marques was presented, together with another project for building railway lines from Inhambane to Transvaal, from Beira to Alto Sanhate and from Lourenço Marques to the border of Mussate. This was a way to attract immigrants to settle in the Zambezi and in Lourenço Marques, preferably in the district of Inhambane and in Maxona.

In the first bulletin of the Geographical Society of Lisbon in 1890 a protest is issued from the society to all academies, societies, institutes and newspapers with whom they share strong relationships.

In response to the appeal, in addition to numerous national institutions and private citizens, the Geographical Society of Madrid in solidarity with its Portuguese counterpart, sent a telegram to all its European counterparts adhering to these claims

> …the protests of Geographical Society of Lisbon against the conduct of England inviting the geographical societies of the entire world to take an equal stand on behalf of the Science of geography and historical rights. (Pinto 2011)

The consequences of domestic politics soon gave results and the Regenerator Party, through the newspaper *Gazeta de Portugal*, was inciting to the opposition of the Progressive Party, in power at the time, using expressions leading to insurrection: "these traitors and cowards who should have their punishment in this world or let justice be made to these people with no further ado." (Pinto 2011)

On January 20, 1890, the Perpetual Secretary of the Directorate of the Geographical Society of Lisbon, Luciano Cordeiro, communicated to the members the following:

> This board has decided in good time and before the authorities of the nation, for the renewal of the vow, so many times formulated by the Geographical Society of Lisbon during the fourteen years of their life, with its persistent work, propaganda and offering consultation that a renovation of politics and colonial administration, and imprinting to each a disciplined and practical boost, would fit tightly with all its terms, to the needs and interests of the traditions from the nation's economy. (Pinto 2011)

Thus, the Geographical Society of Lisbon centralises a comprehensive anti-British protest movement, formulating various motions addressed to the organs of political power as well as several national and international institutions. A national subscription for National Defence of the Country was created:

> whose patriotism was put into evidence, literally filled the hall of the Trindade theatre with citizens from all social classes, answering the call of a handful of patriots, on the memorable night of January 23, 1890, when a major Commission was appointed for promoting a subscription in order to acquire all possible ways of for sea defence of the Metropole and Colonies. (Pinto 2011)

The initial commission was constituted by 259 national members of all beliefs and parties, from the more conservative figure, the Cardinal Patriarch, to the most radical, the Republican Manuel de Arriaga. The chairman of its executive committee was the Marquis of Pomares and the Duke of Palmela was vice-president. Other members of the commission also elected were: Eduardo Abreu; Teófilo Braga; Marquis of Praia and Monforte, Rosa Araújo, Latino Coelho; Sousa Martins, Fernando Palha; Salgado Zenha; Bordalo Pinheiro, Roberto Ivens; Luciano Cordeiro and António Enes. The first session of the Executive Committee was held at the Geographical Society of Lisbon, and Eduardo Abreu was elected secretary.

The then President of the Geographical Society of Lisbon, Bernardino Machado, invites in 1912 the Nobel Prize in Physics Guglielmo Marconi to visit Lisbon. Marconi arrives on 22 May of that year accompanied by the Marquis of Solari, by the director of Marconi's Wireless Telegraph Company Godfrey Isaacs and other two private secretaries. Waiting at the Rossio station were the representatives of the Academy of Sciences of Lisbon Trade Association and Industry Association, various diplomatic legations, namely the Italian, Bernardino Machado himself, the general manager António Maria da Silva and the Officer of the Portuguese Navy Almeida Lima.

The next day, Marconi was received in the Belém Palace by the President Manuel de Arriaga. Next he gave a lecture at the Geographical Society, in which he presented the improvement of the wireless telegraphy system. The Journal of Commerce and Colonies referred:

> The members of the Society of Geography and their families, the government, the Senate, the House of Representatives, the city hall, the Minister of England, Italy, the United States and their staff, colleges, many industrialists, regular schools and colleges were invited. The session was attended by the President of the Republic along with his staff. (*Jornal do Comércio e das Colónias*, May 23, 1912)

In an interview given to the newspaper *O Século,* Marconi commented:

> I will not give any conference in the Geographical Society, and also because I haven't brought the devices that would allow me to fully demonstrate the theories of my presentation. It is a simple visit to this Society and in the event of responding to the most likely questions, I will just refer to the most recently introduced modifications in the wireless telegraphy. (...) It is perhaps an unpretentious lecture that intends to pay tribute to the pioneers of the wireless, the ones who unveiled the electric waves. Then, naturally, I will point out the improvements introduced into my invention after 12 years of experience. (*Século*, May 23, 1912)

In session the Vice President of the Geographical Society, Almeida d' Eça and Almeida Lima, intervened, and who once had been part of the committee responsible for purchasing Marconi equipment on behalf of the Portuguese Navy.

The Geographical Society of Lisbon had many distinguished visitors, such as: Otto Schutt; Verney Lovett Cameron; King Edward VII; Wilhelm II, the German Emperor; President Émile Loubet; Louis Alexander Mountbatten and Sir Clements Markham.

Conclusion

Throughout the period of the *Estado Novo* the Geographical Society of Lisbon maintained its influence over overseas politicies. With the coup of April 25, 1974 and subsequent decolonisation of former "overseas provinces", the institution would go through a troubled period, following the departure of its former president, Professor Adriano Moreira, to Brazil.

In order to keep the Society's main assets and to avoid them being lost during the difficult year of 1975, the management of the Geographical Society of Lisbon was delivered to Admiral Fernando Simões Coelho da Fonseca.

With the normalisation and stabilisation of political conditions that allowed the regular functioning of institutions, the Geographical Society of Lisbon sought to regain its prestige nationally and internationally, succeeding thanks to the persistence of those who never abandoned the basic principles present at the birth of the Geographical Society of Lisbon.

Bibliography

Aires, L. 2001, *Tesouros da Sociedade de Geografia de Lisboa*. Lisbon: Edições Inapa.

Boletim da Sociedade Geografia de Lisboa. n. 1 (1890). Lisbon. Sociedade de Geografia de Lisboa, 9th series.

Bota, A. L. M. G. 1991, "A Sociedade de Geografia de Lisboa: fundação e primeiros anos—Aborda os assuntos de Portugal e África. *Boletim da Sociedade de Geografia de Lisboa,* n. 7-12, 109th series.

Cordeiro, L. 1993, *Questões Coloniais*. Lisbon: Veja.

Estado-Maior do Exército, 1989, *Resenha Histórico-Militar das Campanhas de África (1961-1974)*. Comissão para o Estudo das Campanhas de África. Lisbon: Estado-Maior do Exército, vol. 1.

Graça, J. R. 1867, "Descripção da viagem feita de Loanda com destino ás cabeceiras do rio Sena, ou aonde for mais conveniente pelo interior do continente, de que as tribus são senhores, principiada em 24 de abril de 1843", *Annaes do Conselho Ultramarino. Parte não-oficial.* (1854-58), 1st series, Lisbon: Imprensa Nacional.

Guimarães, M. A. 1984, *Uma Corrente do Colonialismo Português*. Lisbon: Livros Horizonte.

Jornal do Comércio e das Colónias. n.º 17 471 (23 de Maio de 1912). 1

Machado, J. T. M. 1980, "Luciano Cordeiro. Ilustrado", *Boletim da Sociedade de Geografia de Lisboa,* n. 1-3, n. 4-6, 98th series.

Pequito, R. A. 1901, "Como se fundou a Sociedade de Geographia de Lisboa", *Brasil-Portugal*, n. 70, v. 3.

Pinto, R. M. C. 2011, "Das raízes do Mapa Cor de Rosa ao Ultimatum" *Boletim da Sociedade de Geografia de Lisboa*, n. 1-12, 129th series.

Porto, A. F. F. S. 1891, *Silva Porto e Livingstone. (Manuscripto) 1891. Acessível no seu espólio.* Lisbon: Sociedade de Geographia de Lisboa, Typographia da Academia Real das Sciencias.

Século, n. 10 936 (May 23, 1912).

Serra, J. B., 1990, "O Ultimatum visto por Rafael Bordalo Pinheiro", *Publico-Magazine.*

Sociedade de Geografia de Lisboa, 1950, *Sociedade de Geografia de Lisboa: 75 Anos de Actividades ao Serviço da Ciência e da Nação. 1875-1950.* Lisbon: Sociedade de Geografia de Lisboa.

—. 2000, *Sociedade de Geografia de Lisboa 125 anos. 1875-2000.* Lisbon: Sociedade de Geografia de Lisboa.

—. 2002, *Sociedade de Geografia de Lisboa: Estatuto Geral (aprovado pela assembleia geral em sessão de 3 de junho e sancionado por alvará de 3 de julho de 1895).* Lisbon: Sociedade de Geografia de Lisboa.

Unknown Author, 1890, "Expedição para Mantua: diário", *Boletim da Sociedade Geografia de Lisboa.* nº. 8-9.

CHAPTER FIVE

SCIENTIFIC SOCIETIES DURING THE PORTUGUESE FIRST REPUBLIC: AN HISTORICAL PERSPECTIVE ON CANCER AND RADIOACTIVITY STUDIES

RICARDO MOREIRA

The first thirty years of the twentieth century were particularly auspicious for Portuguese science. This fact bears a close relation to the development of already active scientific associations but also to the emergence of new ones. After the republican reforms of 1911, and taking advantage of a growing disposition of the new generation of young intellectuals towards international experiences, the "research school" of Mark Athias (a Portuguese pioneer of biological and medical research) found fertile ground to develop its associative projects and raise the national scientific research standards to higher levels. The Radiology and Oncology fields profited from this dynamic of scientific activity in a way that just a few other fields of research have been able to. Henceforth, they found the means for the implementation of the kind of practices, either in medical research, laboratory work or hospital organisation, that we may nowadays recognise as being of a modern character. The "Radium Pavilion" of the Portuguese Institute of Oncology, that opened in 1933, or the great cancer hospital block, built in Lisbon during the 1940s and which was planned to support the centralisation of the oncological practice at the national scale, are some of the best exemplars of this modern look of the Portuguese science, during the first half of the twentieth century. Scientific societies must have kept pace with these modernising events, and incorporated these novelties in their very statutory body.[1]

[1] The present work is an outcome of the research project *Portugal Nuclear: Física, Tecnologia, Medicina e Ambiente (1910-2010)*, financially supported by FCT - Fundação para a Ciência e a Tecnologia (ref. HC/0063/2009).

The oncological medicine and the medical uses of radiation physics in Portugal

Nuclear Physics originated in the final years of the nineteenth century from the works on radiations of Henri Becquerel (1852-1908), during his research on the relations between phosphorescence and the x-rays, discovered by Wilhelm Roentgen on November 8, 1895. Becquerel noticed, soon after, that a certain uranium ore could produce a kind of radiation identical to the x-rays and that it could penetrate through opaque material bodies while, by itself, impressing photographic plates (Hughes 2002, 352).

In 1897, Pierre Curie (1859-1906) and Marie Curie (1867-1934), began to study this uranium radiation and found a way of quantifying it, after which, in 1898, they discovered two new radioactive elements: polonium and radium (Hughes 2002, 352). The latter, when isolated, seemed capable of a radioactive discharge a million times greater than that of uranium (Curie 1904).

Standing upon those new insights on the physical properties of radiations and radioactive materials and aiming to obtain a deeper understanding of the properties of this new element, the Curies developed a rudimentary method for the fabrication of radium which involved massive quantities of raw materials and reactants.

> In order to obtain one gram of radium bromide ($RaBr_2$) (…) 400 to 500 tons of radioactive ore, 150 to 200 tons of several reactants, 150 to 200 tons of coal, and 800 to 1200 tons of water are needed, which amounts to 1600 to 2200 tons of raw materials, without taking into consideration neither the labour of thirty workers employed in the factory, nor the [work of the] miners that extracted the ore. (Houllevigne 1927, cit. in Pinnel 2002, 35)

Notwithstanding its initial purpose of supplying the chemistry laboratories, this manufacturing was giving rise, even before the years of the Great War, to an important radium industry in France, of which Pierre and Marie Curie, contributing with the necessary human and technical resources, were major associates.[2]

The radium economy had already experienced rapid growth in the first decade of the century (Roqué 1997). By then, Portugal was one of the most important suppliers of uranium ore and participated extensively in

---

[2] On the French radium industry and its relations to the Curies see Boudia 1997; Vincent 1997; Roqué 2001; and especially Roqué 1997.

this industry through the export of autunite (and some torbernite), mostly extracted from the ore deposits of Nelas, in the Viseu region. By means of an industrial production of radium in a factory at Sabugal (Guarda) that had been built in 1908, Portugal had also begun to export radium salts. From 1912 on, the factory was owned by a Franco-Portuguese mining company (Lepierre and Leite 1933). The salts of radium bromide produced there found their way to France through Lisbon, but during their transportation some of it was delayed in the capital in order to be applied in the University Hospital sessions of radiotherapy, after which it supposedly resumed its due route to Europe (Guedes 1916).

Radiations were then increasingly used, mainly in medical applications. If on the one hand, Roentgen discovery had the stunning effect of allowing the medical gaze into the invisible inner part of the living body, without the requirements of the usual surgical procedures–a skill that made possible the new discipline of Radiology–on the other hand, experimental evidence was becoming available on the destructive power of radiations over organic living tissues, the comprehension of which led to the therapeutic use of radioactive emissions as a kind of cautery. Medicine was then one of the most obvious domains where x-rays and radium radioactivity could be applied.

Since 1901, the French physicist Henri Alexandre Danlos (1844-1912), had been delving into the therapeutic uses of radium and applying it to cutaneous lesions, at a time when some of the techniques for the manipulation of the radiations were already known; a fact that enabled laboratory research on the impact of radioactive decay upon the living body.[3]

From 1903 until the Great War, a considerable number of radium and cancer institutes, conducting both medical practice and scientific research, appeared in Europe. This emergence of new scientific institutions and the simultaneous radiological innovation occurring in already existing ones stimulated the radium market, boosting the medical demand for this material. At that time, the price of radium-element exploded, encouraging an increase of its industrial production. Radium had then turned out to be the most expensive substance ever known. Even so, the international availability of radioactive material increased and its medical use became customary. The radiotherapeutic effects (from radium and x-rays alike) against cancer were promising and the number of applications upon

[3] Despite the innumerable researches that the therapeutic usages of radium had motivated, Danlos appears as one of the first to introduce research proceedings with an experimental control of radiations: see Guedes 1921, 503.

patients soared at pace with the creation of the first clinical and research facilities (Costa 2010, 128; Hughes 2002, 354).

For about two decades, until the inter-war years, radiations usage became an inseparable part of cancer therapy. Surgery and Pathological Anatomy were then mature and prestigious medical disciplines. However, when the considerable lack of scientific knowledge on cancer did combine with the auspicious outcomes of the application of radium and x-ray machines on oncological patients, radiologists became crucial subsidiary practitioners in the hospitals' clinics of surgery (Pickstone 2007).

The pioneers of Radiology arose in Portugal promptly after Roentgen's discovery, in the last years of the nineteenth century. Still, apart from a few private practitioners, the employment of radioactive substances in cancer therapeutics had only been definitely introduced in the country through the treatment sessions that took place at the First Surgery Clinic of the University Hospital of Santa Marta (Primeira Clínica Cirúrgica do Hospital Escolar de Santa Marta) in Lisbon, during the year of 1912 (Gentil 1930, 13).

Physicists' research in laboratories and physicians' experimentation in medical contexts evolved side by side. On the one hand, they were trying to understand and achieve a more perfect clinical usage of radioactivity and x-rays; on the other, both were eager to investigate the physical properties of these radiations. Today, we can allege that the disclosure of the Portuguese early twentieth century medicine to laboratory work made a perfect link with the exploratory character of radiological practices and with the urgency of scientific research on cancer.

When, in 1923, the Minister of the Public Instruction, António Sérgio de Sousa (1883-1969), created, by decree, the Portuguese Institute for the Study of Cancer (IPEC–Instituto Português para o Estudo do Cancro) based in Santa Marta, the First Surgery Clinic had already twelve years of clinical and research experience over the ontological diocese under the leadership of Francisco Soares Branco Gentil (1878-1964). The institute was then put under the command of this surgeon, who had been heading that hospital's oncological service since the academic reforms of 1911.[4] One of the main goals of Gentil had been to establish in Portugal a national equivalent of the cancer institutes already operating abroad, permitting the country a delegation in the international union that was being formed at that precise moment. Gentil, who was the main instigator of the official creation of IPEC and of whom it is said to have written its

---

[4] See "Criação do Instituto Português para o Estudo do Cancro", Decree number 9333, *Diário do Governo*, 278, I série, de 29 de Dezembro de 1923.

founding decree himself (Costa 2010, 201), had just taken part in a meeting in Brussels promoted by the Belgian National League Against Cancer; a conference that occurred just about a month before the cancer institute was created in Portugal. Anyway, the cancer consultations in the University Hospital apparently registered a steady increase of incoming patients; a growth paralleled by the number of applications of radiotherapy.

IPEC was then designed according to the centralising model that was being followed by some similar organisations across all of Europe (Costa 2010, 201). It merged, in a single institution, medical specialists from disparate disciplinary fields and integrated them in a team-work practice along with the physicists and the specialised nursing and statistics personnel. This centralised configuration allowed mostly for keeping the three branches of the scientific fight against cancer (teaching, treatment and research) united. It concurrently allotted to a single national centre the considerable financial resources made available by the state and required to buy the minimal amount of radium-element essential to begin with the therapeutic and palliative treatments which, in part, justified the rising influx of patients to the medical consultations.

By the time Francisco Gentil created the cancer consultation service in the University Hospital in 1911, assisted by his fellow colleagues, the radiologist Bénard Guedes (1887-1965) and the physiologist Mark Athias (1875-1946), a novel posture of clinical practice was being promoted both in the academy and in the hospital. It was characterised by a particular close collaboration with the university laboratories and the new institutes of the Medical Faculty committed to the research mission.

Since 1912, Bénard Guedes was employing radium and x-rays as subsidiary instruments in surgery and, in some cases, as primary therapeutic or palliative instruments, so that radiations became one of the most popular medical innovations of the time. Some personal connections of Gentil enabled that a portion of the few grams of radium produced in Portugal between 1913 and 1926 (less than 10 grams) could be temporarily retained in Lisbon in order to be applied to the cancer patients of Santa Marta, being then restored and sent back to their initial destination. It seems evident that the radium employed in the patients of the University Hospital must have served the dual purpose of being at the same time a medical technology of therapeutic usage as well as an instrument of scientific propaganda, in the sense that it bestowed to the medical radiologist the appearance of a professional competence to cure, while it granted a precious support to the surgeon. It is plausible that the introduction of radiological techniques had enabled a much greater arrival of patients to the clinical surgery of Gentil (Raposo 1925).

When in 1915 Francisco Gentil was elected to the position of director of the Medical Faculty of Lisbon and simultaneously nominated as director of the First Surgical Clinic of the University Hospital, this clinic apparently adopted the cancer research and treatment as its speciality. Gentil was then finally able to work out a project of scientific development and put it into practice at the University Hospital, together with his professorship activity at the university, and with the assistance of the radiologists and of the laboratory work of histological analysis, this one under the guidance of Mark Athias. As the number of cancer cases observed in the clinic were rising from year to year, it became mandatory that all the clinical materials should be organised and systemised. Gentil made then the provisions to produce an archive with the clinical observations, the records, the surgical specimens, the plaster pieces, all the documentation in the form of photographs and radiographs "and histological preparations, all conveniently connected in a serial way, in order to be easily used and promptly searched". The archive then set up, together with the Museum of Pathological Anatomy specialised in cancer, came to support research and academic work, serving the production of original scientific work and thus impelling Gentil to the foundation, in 1925, of the institute's own scientific publication: the *Archive of Pathology* (*Arquivo de Patologia*).

The new statutory look of the cancer service in Santa Marta, brought about in 1923–as it took the form of an institution of public character, centralising the fight against cancer in Portugal–constituted a political recognition of the scientific work that had been being done since 1911. IPEC was one of the very few organisations established on a national scale and working with a definite scientific scope. It must have constituted one of the first forms of direct and coordinated planning, designed as a national establishment for public health aid, stretching the medical field out of the hospitals and beyond the clinic rooms and ascribing the accomplishment of medicine's social vocation to the laboratory dominion. Furthermore, laboratory work was set up in coordination with a hospital health assistance activity ruled by rationalised management principles and statistical considerations. Portuguese medicine was then decidedly on the move towards modernity.

The scientific societies

The task of organising a national fight against cancer that had begun in 1911 highlighted the blatant contrast between the national medical sciences and the scientific progress of other nations. Due to the influence of a new generation of physicists (some of them became historically associated with the "medical generation of 1911"), a novel kind of attitude was being felt around the professional demeanour of practitioners, that could now pull together two distinct medical contexts; namely, the field of the clinical work and that of the laboratory space, both merging in a common research stance in order to take clinical advantage of the advancements in knowledge resulting from the laboratory work, but also with the intent of doing research based upon the clinical experience. The conservative approach to medical work, inherited from the nineteenth century, was then being shaken by the critical necessity of setting up, from the start, new scientific institutions that were capable of keeping up with the rising and flourishing of the modern medical and natural sciences, built upon experimentation. These new scientific organisations were a demonstration, and a particularly eloquent one, of the radical transformation that was occurring in the national scientific establishment. The old mind-set had begun to change only during the last decade of the nineteenth century, mainly from the initiative of Mark Athias and Câmara Pestana, both strongly encouraged by some of the topmost figures of the Surgical-Medical School of Lisbon (Escola Médico-Cirúrgica de Lisboa), such as Miguel Bombarda (1851-1901) and Sousa Martins (1843-1897) (Celestino da Costa 1927 and 1946).

Reforming Portuguese medical sciences

Câmara Pestana (1863-1899) was the founder of the Bacteriological Institute, a research unit of the Medical Faculty with an important social mission, which would perform a major role in the eradication of the scourges affecting the urban population as well as in immunology research and vaccine production, contributing extensively to the modernisation of the medical sciences in the capital. Mark Athias, after having studied in Paris during graduation, returned to Portugal and made his way through several institutions, the Câmara Pestana institute included. He left an extremely important scientific legacy and a large number of disciples, being one of the founders of cancer studies in Portugal. Meanwhile, he committed himself to a new synthesis of the French physiological tradition with the German school of Histology, giving rise to a particular approach to the histo-physiological sciences in Portugal (Amaral 2006).

Being responsible for the research section of the IPEC since 1923, Mark Athias devoted himself utterly to the physiology laboratories of the institute. However, before entering this new period in his career, he had become one of the most dynamic and prolific scientists and scholars of his generation, contributing to a great extent to the improvement of the national medical sciences, promoting research, delivering original work, publishing at home and abroad and engendering new scientific societies in the national sphere.

From the outlook of the medical field and considering the new role of scientific societies in the republican Portuguese society, we may come to recognise a significant change that was taking place through the efforts of this generation of physicians. They were performing the modernisation of the medical profession through the intersection of academic work with both a research endeavour and a clinical practice grounded on experience, all guided by the commitment to a sort of mission of social betterment, based on the progress promised by way of medicine and science.[5] We can assess the transformations occurring at the time through the writings of those who, from within the medical field, intended to make the history of those events. One of these figures, and one of the most prominent historians of the Portuguese medicine, was the physician Augusto Celestino da Costa (1884-1956). Also as an element of the renowned medical generation of 1911, he claims that the existing medical scientific association, the Society of Medical Sciences of Lisbon (SCML Sociedade das Ciências Médicas de Lisboa), founded in 1822 during the liberal revolutions, had been, up to that time, together with the Medical-Surgical School and the Hospital, one of the institutions that almost solely made Portuguese medicine. Only then, unable to fulfil the mission of carrying out its own original work and contribute to the scientific progress of the national medical sciences, the Society had merely remained on the observance and acceptance of the transformations that were taking place elsewhere, resigning itself to a role that was "destined to lose its relevance". It could not cope, at last, with the "awakening of the Portuguese medicine to research work", during the last decade of the nineteenth century (Celestino da Costa 1946, 378).

Even after the republican revolution of 1910, the Society of Medical Sciences was simply performing the role of updating and spreading scholarship information in the domains related to clinical practice, never being able to hold a position on the research scene. The scientific production would be, in the medical field, almost fully taken over by the universities and academic institutes. The Society hosted a scientific periodical–the *Journal of the Society of Medical Sciences of Lisbon*–and waged conferences, such as the one of 1907, where Francisco Gentil

[5] On the relations of the modernisation of medicine with the commitment to a political endeavour of social transformation see, for instance, Amaral 2006, Garnel 2010.

promoted a series of lectures on cancer. Despite the importance of these initiatives, it was simply limiting itself to learn from the foreign centres of scientific activity. Its character prevailed as a "mutual education" association (Silva Carvalho 1927, 23); a hoary institution, where the associative body enjoyed a certain prestige, exclusivity and mutual recognition. Just like the Academy of Sciences of Lisbon (see the first chapter of this book), the SCML was not capable of keeping up with the movement of transformation in the national scientific mind-set and…

> at last, it was indeed necessary to leave it, to create other associations with a greater scientific spirit, fitting to the movement that had been set in motion, since the Academy [of Sciences] is also not in the condition to perform that role (Celestino da Costa 1946, 378).

Mark Athias was one of the scientists that felt that urgency. During the politically turbulent years around the formation of the republican regime, the disposition of the new generation for research and for engaging in laboratory work allowed the natural sciences to advance, side by side with the medical knowledge.[6] The influence of medical research practice in the field of the natural sciences became thus evident with the creation of the Portuguese Society of Natural Sciences (Sociedade Portuguesa de Ciências Naturais), in 1907, in which practically a third of the founding members were physicians practising in Lisbon. This society had its main office at the Câmara Pestana Bacteriological Institute–then under the leadership of another recognised figure of the 1911 group, Aníbal Bettencourt (1868-1930)–where the "school" of Athias was a major influence. Athias was then one of the main researchers at the institute, holding a prominent coordinating position as "*chefe de serviço*". Out of the Society of Natural Sciences would emerge, in 1920, the Portuguese Society of Biology. Both were founded under the leadership and scientific guidance of Mark Athias and its "physiology research school" that superintended to the scientific publications (Amaral 2006, 203-209).

---

[6] Celestino da Costa claims that "the revolution made [in the medical sciences] (…), while going to create between us a scientific movement, did not took place in the medicine schools. Whereas reviving an old national tradition, that of the naturalist physicians, that of the influence of our [traditional] medicine in the Natural Sciences, also the emergence between us of the medical scientific research had at once been felt in those sciences (…). The foundation of the Portuguese Society of Natural Sciences in 1907, credited to Athias, and later that of Biology, also from the same initiative, in 1920, [and] the activity of Carlos França as a naturalist was some of the primary agents of this repercussion" (Celestino da Costa 1946, 369-70).

The *Bulletin of the Portuguese Society of Natural Sciences*, put to press in French, granted these young medical researchers and clinicians the opportunity to deliver and disseminate their scientific work abroad, where the *Bulletin* seems to have been received with approval by their peers. In 1908, the Society was already receiving 120 foreign scientific journals and 14 national ones in exchange for the *Bulletin*. By 1942, this deal would amount to a number of 442 different publications that were coming into the Society's library (Athias 1942, x).

Despite the fact that the Society of Natural Sciences was utterly created with the purpose of making provision for the establishment of a Marine Biological Station in Portugal, on the belief that through the collective and coordinated influence of its members it would be possible to obtain from the public administration the necessary funding and conditions that were required for its accomplishment, the Society did not succeed in this matter. Several years were spent in persistent attempts to set up the desired scientific station at the existing Vasco da Gama Aquarium and Museum (Aquário Vasco da Gama) which had been put under the management of the Society. Nevertheless, the efforts were in vain and the station never came to exist (Athias 1942, x).

Yet, other purposes had been accomplished. Mark Athias's point of view, expressed in 1942, stated that it still had been possible to "create an association in which everybody could partake" and consequently to fulfil the desire to develop in Portugal an enduring cultivation of the Natural Sciences, to improve its education, to publish and spread original scientific work produced in the country and to raise an estimable system of exchange of scientific journals (Athias 1942, x).

It seems that the Society had been living until 1919 with some serious treasury problems. However, in that year, it managed to secure an endowment from the public budget, which brought some security, and even a certain leeway, on the commitment to continue with its publishing policy. At that moment, the founding group, closer to Athias "school", must have pondered the convenience of a greater degree of disciplinary specialisation and, as long as the scientific activities of the Society were now guaranteed, they decided that the occasion had come to create an affiliate. Under the patronage of the Société de Biologie (of Paris), the group founded, in 1920, the Réunion Biologique de Lisbonne, which was, two years later, converted in the Portuguese Society of Biology (Athias 1942, x). This new association was by that time putting its scientific work to press via the Société de Biologie, that hosted the articles from the Portuguese counterpart at its own journal, the *Comptes Rendus*, where Athias had been prolifically publishing since 1895 and where the Parisian society put forth its own associates' works as well as those of their subsidiary societies (Athias 1942, x).

The establishing of these two scientific associations in Lisbon may have contributed, without a doubt, to grant the members of the physiology "research school" of Athias an ascendency over other scientific centres, by means of a pervasive control of the editorial activities in the field, a fact that sanctioned them a supervision over most of the national scientific production in the biological sciences (Amaral 2006). Furthermore, these associative initiatives must have contributed to crucial modifications to the conditions in which education and research were assembled in the scientific fields engaged with the medical sciences, with particular important outcomes with regard to the cancer studies.

Along with the professionalisation of the scientific research and clinical work, a growing tendency of specialisation in the medical sub-disciplinary fields would lead, step by step, to an unprecedented proliferation of scientific societies in the years to come, each one of them related to a particular medical occupational domain. The case of radiological practice is remarkable and one of the first examples of this process of disciplinary segmentation.

Technical expertise and risk in professional radiology

During the 1920s, the technical improvements in the employment of radium and x-rays and the increasing reliability of the equipment, enabled the normalisation of radiotherapeutic procedures and the formulation of patterns for therapy, accordingly to the different types of cancer (Costa 2010, 412 and following). The treatment and diagnosis through the use of radiations in the clinical context became an everyday practice and the applications were done at a high pace, allowing for a good level of optimisation of extant resources. From 1927 on, with the new hospital infrastructures of the Portuguese Institute of Oncology (IPO–the former IPEC was now renamed after moving from Santa Marta to the new site at Palhavã, just by the city inner limits), a new unprecedented growth in the clinical usage of radiations became evident, moreover intensified when the new Radium Pavilion became operational in 1934.

The normalisation that emerged from the established routine of clinical procedures, brought with it the specialisation in particular hospital functions and the division of labour, with the consequent outcome of the fragmentation of practical and theoretical knowledge; a partition in the medical act that denoted a rationalisation principle concerning hospital work (Costa 2010, 256).

The decision of centralising the fight against cancer in Lisbon and of entrusting to a single institution the functions of teaching, researching and treating at the national scale, may have, without a doubt, heightened the already existing tendency towards the fragmentation of the hospital work at IPO. It intensified the breaking down of the clinical activity in its various separate moments and subdivided the medical work in its many sub-disciplines and subsidiary medical domains, forming a work context much more favourable to the segmentation and to the constitution of scientific or professional associations (Costa 2010, 464).

By the time operational routines at IPO assumed the traits of a sort of "factory of cure" (Costa 2010, 256), the labour division in the hospital working pattern and its intensifying propensity for partition must have provoked the emergence, with greater acuity, of the risks immanent to each of the operational tasks, at each of the sections of the institute. Especially the radiologists must have felt, in the highest degree and in their own flesh, during their customary practices of diagnosis and treatment, the consequences of this rationalising model to which they were committed without the required protection apparatus against radiations.

In 1930, after some blood tests on technicians and physicians had exposed a few cases of serious anaemic disorder, the director Francisco Gentil suspended the radium treatments (Gentil 1933). In 1934 radium applications were resumed, this time being following the required safety measures, convened in an pioneering architectural project by Carlos Chambers Ramos (1897-1969) that became widely known as the Radium Pavilion. This was one of the first buildings, surely the first in Portugal, paying due attention to the security conventions put forth in the International Congress of Radiology of 1928 in Stockholm (Gentil 1933).

Founded in 1931 by a group of distinguished radiologists–among whom could be noted the presence of Bénard Guedes, in charge of the radiology section of the IPO as well as of its treasury department–the Portuguese Society of Medical Radiology (SPRM–Sociedade Portuguesa de Radiologia Médica) adopted from the outset the juridical personality of a professional association ("*associação de classe*"). It had as one of its central purposes "to study and disseminate the measures for individual and collective protection, required to avoid the harmful effects and the risk of mischance that may arise from the use of x-rays and radium and similar radiations".[7]

In contrast with the scientific societies hatched by Mark Athias, along with his colleagues and disciples, around general sciences that convoked a

[7] See "Estatutos da Sociedade Portuguesa de Radiologia Médica", 1931.

wide array of professional skills and qualifications, the SPRM formed a real professional organisation. At first it was presented to official ratification as a "scientific association", but the public officer suggested, without further explanations, to change the denomination, probably in order to fit the new corporative ideology promoted in public administration.[8] Besides an expected scientific concern with particular questions of the radiological expertise, this association looked as if fashioned to act primarily as an instrument for the enhancement of a professional group and for the care of its collective concerns. The other purposes of the association were "to instigate the study of Radiology" and "to promote the solution of questions that matter to the professional life of all of those who are devoted to Radiology".[9] Not included in these statutory norms were those kind of philosophical concerns that we would predictably find in the statutes of the Society of Natural Sciences, configured by such purposes as were the "cult and development of Science", or the incitement for publication of original research work. These were goals denoting a certain ideal of what the scientific practice should be and that partook of something like a "social vocation"[10] of the medical endeavour, fitting a militant positivism and a confidence in the power of reason to illuminate–together with the laboratory work and the reliance in experience and facts–the conception of a new model of justice, in harmony with the particular conditions of each society. Those scientists that first conjoined with the new republican regime and promoted the republic, believed, as was common in Europe, in the power of science to lead to progress and civilisation and to guarantee the welfare of humankind.[11] Therefore, they carved out their science in the image of that idea.

---

[8] *Ibidem.*

[9] *ibidem*

[10] The expression "social vocation" is used by Maria Rita Garnel to refer a characteristic particular to the turning of the century medicine commitment to deal with the social scourges affecting large sectors of the population. This posture went along with an engagement in political activities in order to secure social prestige and public commitment to the medical social mission. See Garnel 2007, Amaral 2006, 177 and following.

[11] Miguel Bombarda was one of the most paradigmatic representatives of this optimistic scientism, combining it with his republican militancy. He put forth his philosophical thinking in a work of 1897 entitled *A Consiência e o Livre Arbítrio* (*Conscience and Free Will*).

Conclusion

After the discoveries of the x-rays and radioactivity and of the means to effectively manipulate these new forces, medicine adopted radiological practices as a new promising method of diagnosis and treatment. Cancer, emerging at that precise moment as one of the most pressing problems of medicine, was assigned almost immediately to the new radiological endeavour and cancer studies developed, virtually from the beginning, in close relationship with the exploration of the physical properties of radiations. Radiation physics and medicine progressed then in close collaboration with one another.

Simultaneously with the emergence of medical radiology, research work and laboratory practices were introduced in Portugal by the hand of a few medical scientists and physicians who became widely associated with the republican reforms of 1911. These reforms were designed to enhance the development of the new posture of scientific work and created the optimal conditions for the experimental sciences to become the axial feature of a novel medicine being promoted in a recently founded university, with a new medical school and a renovated hospital. The conditions of practice suddenly became sufficient for the emergence of new disciplines and institutions. The development of cancer research and hospital therapeutic sessions, as well as the work being done by the histo-physiological practitioners and radiological specialists along with traditional medical disciplines such as Pathological Anatomy and Clinical Surgery, led to the formation of the Portuguese Institute for the Study of Cancer. In due time, this would become the most expensive scientific medical institution in Portugal.

At the same time that medical research became customary in the institutes of the faculty and an important part of the graduation programme, new scientific societies were rapidly occupying the vacant places left by the old existing ones. The fact that scientific medical practice had never been much concerned with research or with the production of original work meant for these new generations of practitioners that all was left for them to do. The Society of Medical Sciences of Lisbon never committed to the mission of producing new medical knowledge and did not even create the proper conditions for such work to be done. The new Portuguese Society of Natural Sciences and the Portuguese Society of Biology, created in 1907 and 1920 respectively, assumed that role and in a few years established a fruitful scientific activity, promoting the association of new members, the publication of scientific journals for the dissemination of original work and the exchange of those publications in a movement of internationalisation that proved to be particularly successful and allowed for the creation of a significant specialised scientific library. These new societies made possible the intersection of academic activity with the research practice and prepared the way for the consecration of the leading role of the histo-physiological "school" of Mark Athias at a national level.

When the medical work became increasingly devised by rational and systematic principles of organisation, with the scientific work being done in close relation to the clinical practice and at a high pace of activity, the configuration of the new hospital infrastructures of the cancer institute conveyed the conditions to fulfil the plan of implementing a "factory of cure" for cancer patients. This led to a tendency towards specialisation and to the increase of work intensity during the medical procedures in each occupational task. Radiologists felt this growth of intensity at work in a most painful way, for they were daily exposed to radiation without observing the necessary protection rules. This led to two major events in the radiology field. The first was the construction in 1931 of the Radium Pavilion of the Institute of Oncology, in observation of the safety norms approved in 1928. The second was the creation of the Portuguese Society of Medical Radiology, in 1931. This was a scientific association with a professional character, and in that aspect it was different from the preceding ones. Not so much concerned with a proper scientific *ethos* or with the development of science itself, this association was mainly interested in the promotion of the profession in the public sphere, and with the protection of the professional concerns of its members, especially regarding the implementation of the necessary security practices against radiations.

Radiology had, by that time, become a consolidated discipline with a proper place in the medical field. Laboratory experimentation and scientific questioning had then given rise to much of the routine practice and technical expertise. The great expectations of social transformation that were behind some of the turn of the century scientific enterprises were by the mid-thirties partly accomplished and enmeshed with the rationalisation and coordination of the hospital ordinary working, based on an optimised management of resources and statistical guidance. The emergence of new scientific professional associations must have then reflected the transformations taking place in everyday practice, in hospitals and research institutions alike.

Bibliography

Amaral, I., 2006, *A Emergência da Bioquímica em Portugal: As Escolas de Investigação de Marck Athias e de Kurt Jacobsohn*, Lisbon: Fundação Calouste Gulbenkian.

Athias, M. 1942. "A Sociedade Portuguesa de Ciências Naturais", in *Actas do 1º Congresso Nacional de Ciências Naturais*.

Bombarda, M. (1897) 1902, *A Consciência e o Livre Arbítrio*, Lisbon, Livraria Editora.

Boudia, S. 1997, "The Curie laboratory: radioactivity and metrology", *History and Technology*, 13, 249-265.

Costa, R. 2010, *Luta contra o cancro e oncologia em Portugal. Estruturação e normalização de uma área científica (1839-1974)*, Porto: Faculdade de Letras da Universidade do Porto (PhD thesis).

Celestino da Costa, A., 1927, "A Investigação Científica na Sociedade de Ciências Médicas de Lisboa", *Sociedade de Sciências Médicas de Lisboa (1822-1922), comemoração centenária em 1923*, Coimbra: Imprensa da Universidade.

—. 1946, "A evolução das ciências médicas em Portugal", *Jornal da Sociedade de Ciências Médicas de Lisboa*, 8, 363-383.

Curie, M. S., 1904, "Radium and Radioactivity", *Century* Magazine, 461-466.

Garnel, M. R. 2007, *Vítimas e Violências na Lisboa da I República*, Coimbra: Imprensa da Universidade de Coimbra.

—. 2010, "Corpo: Estado Medicina e Sociedade no Tempo da Primeira República", In *Corpo: Estado Medicina e Sociedade no Tempo da Primeira República*, Lisbon: INCM, 7-11 (Catalogue of the homonymous exhibition, celebrating the 1st centenary of the republic).

Gentil, F. 1930, "O Instituto Português do Cancro" in *Arquivo de Patologia*, 2.

—. 1933, "O Pavilhão do Rádio no Instituto Português de Oncologia", *A Medicina Contemporânea*, 261-264.

Guedes, F. B. 1916, "O rádio no tratamento das neoplasias" *in Portugal Médico*, n.º 2.

—. 1921, "Sobre o tratamento do cancro", *Portugal Médico*, 12.

Hughes, J. 2002, "Radioactivity and Nuclear Physics", In *The Modern Physical and Mathematical Sciences*, ed. M. J. Nye, Cambridge University Press.

Lepierre, C. & A. P. Leite, 1933, *Uma indústria de origem francesa: A indústria do Rádio em Portugal*, Coimbra: Publicações do Instituto Francês de Portugal.

Pickstone, J. 2007, "Contested Cumulations: Configurations of Cancer Treatments through the Twentieth century", *Bulletin of the History of Medicine,* 81, 164-196.

Pinnel, P. 2002, *The Fight Against Cancer: France 1890-1940,* London: Routledge.

Raposo, L. S. 1925, "O Instituto Português para o Estudo do Cancro", In *O Ensino Médico em Lisboa: A Patologia e a Terapêutica Cirúrgicas*, ed. F. Gentil, Lisbon: Faculdade de Medicina de Lisboa.

Roqué, X. 1997, "Marie Curie and the radium industry: a preliminary sketch", *History and technology*, 13, 267-291.

—. 2001, "Displacing radioactivity", in *Instrumentation between Science, State and Industry*, ed. B. Joerges and T, Shinn, Dordrecht: Kluwer Academic Publishers.

Silva Carvalho, A., 1927, "A Primeira Sociedade de Ciências Médicas de Lisboa", *Sociedade de Ciências Médicas de Lisboa (1822-1922) Comemoração centenária em 1923*, Coimbra: Imprensa da Universidade.

Vincent, B. 1997, "Genesis of the Pavillon Pasteur of the Institut du Radium of Paris", *History and Technology*, 13, 293-305.

CHAPTER SIX

MODERNISING AMBITIONS: AGRONOMISTS IN ACTION BETWEEN DICTATORSHIP AND DEMOCRACY (PORTUGAL, 1957-1986)

DULCE FREIRE

The Centro de Estudos de Economia Agrária (CEEA) was created in 1957. This scientific research unit, sponsored by the Fundação Calouste Gulbenkian (FCG)[1], was active for almost 30 years. It operated during the decades in which major changes occurred in agriculture and rural society in Western Europe and, especially, in Portugal. Since the Centre was active until 1986, its researchers accompanied the major changes of the transition from dictatorship to democracy (after the 1974 Revolution) and the preparation of the Portuguese adhesion to the European Economic Community (1986). During these decades, different factors contributed to maintain agriculture on the core of national and international debates. In this context, this chapter analyses the origins and the main objectives of the CEEA, evaluating its role in the discussions associated with agricultural modernisation.[2]

---

[1] In this article, the names of the institutions are in Portuguese, Centro de Estudos de Economia Agrária (Centre for the Study of Agricultural Economics) and the Fundação Calouste Gulbenkian (Calouste Gulbenkian Foundation).

[2] Research for this article was developed with the support of a post-doctoral scholarship (2008-2011) funded by the Fundação para a Ciência e Tecnologia (Foundation for Science and Technology) and the research project "Portuguese agriculture: food, sustainability and development (1870-2010)", which has been developed (2012-2015) at the Instituto de Ciências Sociais da Universidade de Lisboa (Institute of Social Sciences, University of Lisbon) and funded by the same foundation (www.ruralportugal.ics.ul.pt).

In this analysis, two main questions are assessed. One is related to the conditions of development of scientific research and its application to Portuguese agriculture during the decades that followed the end of World War II. This issue has not yet been fully elucidated by Portuguese historiography. Although it is recognised that during the 1950s there was investment in agronomic research, particularly during the execution of the Marshall Plan (Rollo, 2007), there is little evidence of the impact of innovation in agriculture in this decade and the next (Freire, 2007). This is an essential topic to understand the historical circumstances of the diffusion of the *green revolution* model in Portugal. Despite the criticism that the *green revolution* has been subjected to in recent decades (Griffin, 1979; Glaeser, 1987; Lanero and Freire, 2011), it remains relevant to explain how the model has spread throughout the world, including the outlying regions of the Iberian Peninsula (Lanero and Freire, 2011b). Knowledge on the activities performed by the CEEA is a contribution to shed light on the Portuguese case.

Another goal of this analysis is to evaluate how political regimes conditioned the actions of civil society. This is a big issue in the international debate, which obviously cannot be fully treated in this chapter. However, to understand the origin and development of the CEEA it is necessary to take into account the complex relationship established between the State and technical elites. In the second half of the twentieth century, the insertion of Portuguese agronomists in epistemic communities was facilitated by the admission of the country in major international organisations.[3] Nevertheless, the dictatorship continued to limit the internal action of these elites and conditioned the creation of professional associations (Delicado *et al.*, 2012). The CEEA was an exception. Much as has happened with other organisations, the CEEA was ambiguous, showing similarities both with the think tanks and with the academic centres for applied research arising in Europe during these decades (Stone and Denham, 2004). Therefore, the actions promoted by the CEEA need to be placed in the broader context of political and social changes that were taking place in Portugal and in the world during its 30 years of existence.

The chapter is divided into two parts. First it describes the origin and the consolidation process of the CEEA, which requires connecting the Centre to the complex networks which link the authoritarian state, civil

[3] Although it continues to be ruled by a dictatorship, Portugal was admitted to the major international organisations related to economic and social development created from the 40s, namely the Food and Agriculture Organisation (FAO) in 1946 and the Organisation for European Economic Co-operation (first OEEC and later OECD) in 1948.

society and agronomic research. Second, it identifies the objectives that guided the activities of the CEEA, examining how these were articulated with the political and social changes that occurred in Portugal between 1957 and 1986.

Origins of the Centro de Estudos de Economia Agrária

The Centro de Estudos de Economia Agrária had, from the beginning, a unique status in the national scenario of research and debate on agriculture and rural issues. First, because it was part of the Fundação Calouste Gulbenkian. The foundation, based in Lisbon, was created in 1956 to fulfill the testamentary dispositions of the millionaire Calouste Gulbenkian. The institutionalisation of this foundation was long and complex, in part due to the negotiations with the Estado Novo.[4] The dictatorial regime of fascist inspiration, which ruled Portugal between 1933 and 1974, closed numerous associations and repressed civil society action. The Fundação Calouste Gulbenkian was one of the few non-governmental organisations created during the dictatorship, and became one of the main sponsors of artistic, educational and scientific programmes in Portugal.

During the first decades of its activity, the Fundação Calouste Gulbenkian established several scientific research units in areas that were not under state control. The CEEA was the first of these research units. The initiative to set up the Centre came from a group of agronomists who, on September 1957, sent the proposal to the Foundation. In the document, the relevance of the Centre is justified by the urgent need to develop agriculture and increase the incomes of Portuguese farmers. In this sense, the document states that the Centre should be devoted to the study of the following problems: "living conditions of rural populations; profitability analysis of various forms of land use; defining the characteristics of the farming methods most appropriate to the demands of contemporary life; scientific analysis of the social-agricultural income distribution; study of the forms of rural property exploration" (FCG Historical Archive, Pt:93, 22/5).

In September 1957, the foundation President, the lawyer José de Azeredo Perdigão, received nearly twenty proposals with several projects and requests. These requests were for grants and other type of support (to establish research units, to purchase equipment, to support fieldwork, etc.),

[4] Several books gave more information about Calouste Gulbenkian and the foundation (Barreto, 2007; Hewins, 2009; FCG, 2010, for example).

which reflects the multiple expectations placed by the Portuguese elites on this foundation. The decisions of the President were not always favourable, but he had no doubts in accepting the proposal to create the CEEA. He considered that the project "has all the conditions to be accepted and carried out by the Calouste Gulbenkian Foundation" (FCG Historical Archive, Pt:93, 22/5). In the report he sent to the Foundation Board of Directors, he stated that the Centre could contribute decisively "to value our greatest asset – the land" (*ibidem*), so it should be issued "a favourable decision which would allow the proposal to be considered and financed" (*ibidem*). These recommendations appear to have been decisive for the creation of the CEEA for two main reasons. First, Azeredo Perdigão had a strong influence over all the foundation bodies. Second, because both the programme activities planned to be held by the Centre and the agronomist who presented the proposal did not fostered easy consensus among the members of the Board of Directors.

José de Azeredo Perdigão felt himself the need to provide additional information to clarify his position, as far as the goals of the Centre were concerned. Strengthening his opinion became necessary also because, in the early years of the twentieth century, he had published a small essay in which he underrated agriculture and argued that Portugal should focus on industrialisation. However, in 1957, despite the progress of the country's industrialisation after the World War II, agriculture still had a strong economic and social role. Actually, until the early 1960s, the primary sector was the main contributor to the Gross Domestic Product and employed a large part of the workforce (Lains, 2009). Recognizing this Portuguese specificity in the Western Europe scenario, José de Azeredo Perdigão understood that it was necessary to promote the modernisation of the primary sector, allowing it to contribute to the overall development of the country.

The President of the Fundação Calouste Gulbenkian argued that "our agriculture is, even today, guided by the same principles that prevailed centuries ago without taking advantage of the numerous advantages modern science makes available to the farmer, in order to reduce the risks of land exploration and to increase its productivity" (FCG Historical Archive, Pt:93, 22/5). In fact, this opinion provides a sample of the concepts that at the time were common among technical elites in Portugal, and which were used to explain the problems identified in agriculture: outdated practices, low productivity, widespread ignorance among farmers. Although historiography has been demonstrating that these opinions reflected neither the economic behaviour of farmers, nor the technological transformations which were occurring in Portuguese agriculture in the first half of the twentieth century (Brito, Baptista and Pereira, 1996; Freire, 2007), these were the concepts which framed the creation of the CEEA. The Centre took on the mission of studying the best strategies to disseminate new agricultural practices, based on scientific knowledge and modern technology in order to contribute decisively to modernise Portuguese agriculture.

The proposal that led to the CEEA institutionalisation was signed by Mário de Azevedo Gomes and Henrique de Barros, two agronomists recognised by their professional skills and by the political positions they defended, often against the dictatorship. Both had been persecuted due to their active involvement in opposition movements against the Estado Novo and the political police had their criminal records. Among other activities, Mário de Azevedo Gomes had been president of the Movimento de Unidade Democrática, created after World War II to promote the democratisation of the country, but made illegal in 1946. Henrique de Barros was part of several movements that acted on the edge of legality, which did not benefited his academic career in public universities. Several factors contributed for selection of these two names as the official proponents of the document submitted to the Fundação Calouste Gulbenkian. Especially, Mário de Azevedo Gomes had a close personal and ideological relationship with José de Azeredo Perdigão. Both were Republicans and had affinities since the beginning of the *Seara Nova* movement, in the 1920s, before the establishment of the dictatorship. However, the role of Mário de Azevedo Gomes also had a symbolic meaning. On the one hand, he had advocated for long the establishment of a technical assistance service to farmers. On the other, the Estado Novo had extinguished the public research centre Estação Agrária Central, in 1936, when he was its director.

Although these were the protagonists, the information gathered indicates that it was Manuel Bobone who had the idea of establishing a research centre devoted to agriculture. As a public servant in the Direcção Geral de Agricultura (Ministério da Economia/Economy Ministry), this agronomist had visited the USA in the early 1950s. He thus knew the programmes of agricultural research and extension that existed in that country. This initial group was also composed by Quartim Graça and Eugénio de Castro Caldas. The first was a civil servant who had already been the Under Secretary of State for Agriculture in 1947, and again in 1958. The second one, besides being a professor at the Instituto Superior de Agronomia (at the time the sole Portuguese agronomic university), had several political appointment positions (including being a member of several advisory and technical state committees). Even agronomists who were civil servants acknowledged that the dictatorship imposed institutional and political constraints that did not allow the development of technical assistance programmes within the State, either in universities or in other public services. Therefore, some agronomists tried to find other ways to study problems and to promote agricultural innovation. The CEEA was an opportunity to accomplish this purpose. The project gathered agronomic engineers from different generations and ideological positions (who did and did not support the Estado Novo). However, all were renowned and experienced, which gave the project a national goal and withdrew from its immediate political connotations.

The CEEA began operating in January 1958. It followed one of the general requirements of the Fundação Calouste Gulbenkian, which was not to compete with the state, but to act only in areas outside the state's intervention. The CEEA proponents considered that the foundation contributed to overcome an institutional gap, which existed since the 1930s, when the Estação Agronómica Nacional (1936), another kind of state research centre, replaced the Estação Agrária Central. More than a simple name change, this replacement revealed a clear preference of the dictatorship for the agronomy branch related to genetic and fitopathologic research. Thus, it disappeared from the organisational structure of the state one service devoted to the agronomy branch, which studied the economic and social factors associated with the modernisation of agriculture. After World War II, the rapid spread of the *green revolution* gave further relevance to these issues, and in several European countries, they became areas of study in rural sociology. Nevertheless, according to Henrique de Barros, in the 1950s Portugal, this study domain was being "almost marginalised by the people involved in agricultural policy, from farmers to the final decision-makers, including the technicians and the researchers" (Barros *et al.*, 1983). The CEEA proposal was aimed at solving these shortcomings, by actively contributing to study the implementation of *green revolution* solutions in Portuguese agriculture, since these solutions were increasing the productivity of land and labour in other countries.

However, the nature and goals of the CEEA raised prolonged objections, both within the foundation, and in political circles. In an attempt to clear all doubts, José de Azeredo Perdigão provided further clarifications in the first major event of public projection of the CEEA, held in 1959.[5] He reaffirmed that the Centre was not intended to compete

[5] It was the first *Curso de gestão da empresa agrícola*, held on the premises of the Associação Central de Agricultura Portuguesa, in April, 1959. Public and corporatist services officers were present at the inaugural session (FCG/CEEA, 1960:7).

with the state, but rather to work together with all the existing corporate and public entities. Especially because "knowing what to produce, where to produce and how to produce is presently a common concern of governments and farmers" (Barros *et al.*, 1960: 22). For these reasons, across the world, state entities and private initiatives had arisen, joining farmers and technicians and promoting the necessary studies to support the best decisions of the state and agricultural entrepreneurs. Offering to cooperate with the state, the CEEA positioned itself in this lineage of these entities[6] and gave the dictatorship the political decision to accept or not this collaboration to improve public policy.

In fact, during the first decade of the CEEA activities, which coincided with the last years of Oliveira Salazar's government (1932-1968), the criticism of the agronomists regarding the indifference of the State increased. Although the dictator allowed the establishment of the Centre, he was not receptive to the adoption of the modernising solutions advocated by its researchers. The situation changed a little during the Marcelo Caetano period (1968-1974). There was a greater receptivity towards the initiatives and proposals of agronomists, which, however, was not reflected on political measures for agriculture modernisation. Following the principles of the Calouste Gulbenkian Foundation, during the dictatorship, the CEEA activities were based on a plurality of opinions and ideological positions, which guaranteed its independence. However, these principles do not seem to have favoured the Centre's intervention in public policies and it seems to have had only a modest capacity to influence the course of agriculture modernisation (Freire, 2002).

In the years that followed the Revolution of April 1974, the CEEA members had more opportunities to influence public policies. Some of them abandoned temporary the CEEA to assume executive government positions[7] or to work in other public and private organisations (cooperatives, associations, etc.). In this way, they participated actively in the processes devoted to producing fast changes in rural areas or to increase food production.

---

[6] The CEEA members knew some of them, in Europe, the first entities of official nature arose in Germany (1892) and Switzerland (1901). The development of research in the areas of agricultural economy led to the emergence of more public and private centres both in Europe (Netherlands, 1930, France, 1937), and in North America (Caldas and Pereira 1959: 63 - 67).

[7] Among the members and employees of the CEEA who assumed positions of minister and secretaries of state related to agriculture, we may mention Agostinho de Carvalho, António Lopes Cardoso, Fernando Oliveira Baptista, Joaquim Lourenço e João Mendes Espada.

The political and institutional changes introduced by the Portuguese democratic governments and the intensification of international critics to the *green revolution* may have contributed to diminish the relevance of the CEEA project. At the same time, following a strategy to reduce the direct responsibilities in scientific research, the Fundação Calouste Gulbenkian closed several centres formed in previous decades. The CEEA was formally extinct in 1986. However, many of its former members continued to develop professional activities (teaching, research, public employees) related to agriculture and rural issues. After all, these agronomists were part of the technical elite that took on broad responsibilities in national and international institutions.

Activities of the CEEA: to research, to disseminate and to influence

Both the justifications provided by agronomists in the proposal submitted to the Fundação Calouste Gulbenkian and the opinions of José Azeredo Perdigão made clear that, despite the industrialisation processes that were occurring in the country during the 1950s and 1960s, agriculture continued to have an important role in the economy and society. Giving meaning to these perspectives, the new centre would have as a primary goal the development of the agricultural sector, which was seen as immersed in a kind of historical backwardness. It should also contribute to the introduction and diffusion of scientific knowledge and technology that were prosperous in other countries. As it is referred by Eugénio de Castro Caldas and Mário Pereira, in the foreword of the first book published in the CEEA collection, the foreign bibliography that reached Portugal showed that even the Old European Mediterranean countries had decided to adopt the economic research techniques applied to agriculture in the USA (Caldas and Pereira, 1959: 13-14). The CEEA members accepted this mission and presented themselves as active agents in the diffusion of the American agriculture and food paradigm (based on fossil energy, chemicals and genetic manipulation), which became hegemonic in the West after World War II.

Taking into account the social and agro-ecological specificities of the country, the CEEA researchers devised a model aimed at making Portuguese agriculture more efficient. This was, after all, an ancient ambition, probably dating back to the eighteenth century, and lengthily discussed. However, as Azeredo Perdigão carefully points out, in 1960, the purpose was not to reedit physiocrat proposals or anachronistic modernisation programmes (Barros *et al.*, 1960: 17-22). On the contrary, it was to follow the latest international trends. This was the era in that the miracles of improved seeds and chemicals were believable (Griffin, 1978). Agronomists "came out of the schools with a mission to suit the agricultural and farming world to the needs of good techniques, which, by definition, were the most modern ones" (Baptista, 2001: 174). This ambition implied the recognition of the backwardness signs in Portuguese agriculture and their elimination, by following a previously outlined model.

Portuguese agronomists understood that after the State promotion of extensive exploration natural resources, which led to the extinction of uncultivated lands,[8] the place of agriculture in economic development would be achieved through agricultural intensification. These agronomists considered that the general development of the country demanded that agriculture maintained its historical economic and social functions. They considered that, at least in a transitory phase, measures which would require a sudden rural population exodus would have negative effects on society and the economy. Therefore, modern agricultural practices should be adopted, if they could assure the maximisation of natural resources exploration, but also if they ensured the dynamism and well-being of rural society. These perspectives contradicted the economists' position, which advocated a drastic and sudden reduction of the agrarian active population (to a value closer to the 5% of Northern Europe countries). Nevertheless, these agronomists ensured that it was possible to increase the efficiency of farms while maintaining relatively high levels of rural employment.

Members of the CEEA focused on production conditions of goods to satisfy the immediate needs of the growing urban population, which was increasing demand for livestock, fruits and vegetable products. This focus was a high priority because the Portuguese population continued to show calorie and protein consumption levels that were lower than the levels recommended by international organisations (Freire, 2011).

Similar to the large modernisation projects that were being implemented by Franco in Spain (Freire and Lanero, 2013) and by the governments of other countries (Scott, 1998), the plans designed by the Portuguese technicians gave the State a primary role. The modernisation proposals were based on extending irrigation systems, associating them to the multiplication of medium-sized farms focused on market production. Both the expansion of irrigated farmlands and the changes in the agrarian

[8] Two policies promoted by the Estado Novo contributed strongly to reach the limit of extensive land exploration. One was the Campanha do Trigo (Wheat Campaign) in the 1930s (Freire, 2008), the other was the forestation of wastelands in the 1950s (Freire, 2004).

structure required solid financing and public policies. Although the irrigated farmlands could be an incentive for property division, without state intervention this process would be too slow and disorganised. Therefore, in a country with a historical imbalance of land ownership, between the North (small size properties) and the South (large estates), agronomists advocated that public investment in irrigation infrastructures should be linked to a process of property redistribution, that would allow the formation of farms with technically efficient dimensions. In fact, these were old issues, dating back to the nineteenth century, but the national polemics were being updated by international experiences and debates. In short, we can say that the modernisation model proposed by CEEA agronomists was based on intensive and specialised agriculture, developed within technically efficient family farms, thus maximising national production in order to satisfy the demands of internal urban markets.

Although, at the time, the profound complexity of agricultural innovations could not be fully recognised by technicians, it was understood that meeting the demands of an industrialised and urbanised society posed new problems to the farmers. These problems could be divided into two categories. Some problems had exogenous origins, but affected the farmers work and incomes. Amongst these were: the impacts of economic policies; changes in consumer demands; market fluctuations for agricultural inputs and outputs; action of powerful monopolist commercial or industrial organisations. Other problems were indigenous to the farms and were related to their organisation and performance. Among these can be mentioned: extent and purposes of the agrarian exploration; farming systems; productivity of production factors; priority of financial investments; combination of production factors; analysis of farm incomes; production costs.

The CEEA was built to actively participate in solving the endogenous problems and also some of the other problems, through scientific research and dissemination of technical innovations. From the beginning, the mission of the Centre developed on two main fronts. As Castro Caldas and Mário Pereira stress it, in 1959, on the one hand, it was necessary to provide working conditions for Portuguese researchers. The world scientific research system was changing and those who "were aware of the progress of agriculture economics research noticed with anxiety, grief and dismay that everything newly presented was no longer the product of isolated researchers work [...], but the result of systematic research conducted in good economics and rural sociology laboratories, properly organised and with access to means of investigation that did not yet exist in Portugal" (Caldas and Pereira, 1959:14). The Fundação Calouste

Gulbenkian provided the Centre with logistical and financial conditions, which allowed the recruitment of several researchers, technicians, interns and other workers.[9] Following the settings of international academic laboratories, the researchers organised themselves into groups in order to develop research projects, which would require fieldwork and travelling abroad. At the same time, they participated in international conferences, organised several scientific meetings and published research results. They also promoted exchanges with national partners (central and regional public services, universities, corporatist organisations, associations) and international partners (FAO, OECE/OCDE, EFTA universities and public services related to development).

On the other hand, the Centre researchers had to find ways of transferring new knowledge to those who worked the land: "in countries where the economic and social research in agriculture progressed, the farmers professional organisations showed greater vitality and established with the scientific institutions the necessary alliances to achieve common goals" (Caldas and Pereira, 1959:14). These goals will remain throughout the period of the Centre's existence, despite the adaptations made to the various political and social dynamics that occurred during the decades of its activities.

Initially, the strategies adopted to disseminate new knowledge and practices were aimed directly at both technicians and farmers. In this sense, one of the first initiatives of the CEEA was the organisation (in 1959) of a training course that brought together one hundred technicians. In addition to other courses, they also promoted conferences, meetings, and the publication of three simultaneous collections (books, articles and working papers).[10] These publications aimed at disseminating the work of the CEEA members and at the translation into Portuguese of papers or articles written by renowned foreign researchers. With these collections, which were active for more than 25 years, the Centre gathered extensive knowledge on agriculture and rural society, influencing the training of several generations of Portuguese technicians and agronomists.

The direct action aimed at farmers did not have this kind of continuity. It began by giving priority to issues related to agricultural management. It

---

[9] Many of them that could benefit from the scholarship and internship programmes of the CEEA, were recruited among senior students of the agronomic engineering of the Instituto Superior de Agronomia.

[10] The CEEA library was closed and dispersed after 1986, but most of the books that were published in the collections of the centre are available in the current library of the Fundação Calouste Gulbenkian (www.biblarte.gulbenkian.pt) and also in other libraries in the country and in the world.

was precisely to observe the activities of these farmers advising organisations that the CEEA researchers made their first visits to France. The synthesis of what they had learned constituted the first book published in the CEEA collections (Caldas and Pereira, 1959). Until 1958, the agricultural enterprise management “was an unknown theme” (Galvão, 1978: 14) among Portuguese farmers. To make this practice effective, they thought that was essential to create *management advice for the farmers*. The CEEA sponsored four initiatives in different regions of the country (Baixo Alentejo, Beira Litoral, Minho and Douro Litoral), expecting the multiplication of these farmers-entrepreneurs associations. In 1958, the beginning was auspicious, but the expectations were not fulfilled. However, these centres did operate for a few years and promoted technical training for farmers, especially in the north and centre of the country.

The activities that may be considered of rural extension were more visible during the first two decades of the CEEA functioning. Covering roughly the 1960 and 1970 decades, they corresponded to the period of the Centre’s greatest activity and coincided with the peak of the *green revolution* global diffusion. Although the Estado Novo was promoting some changes in agriculture and rural areas (for example, the allocation of subsidies for some farm infrastructure, since 1946; the release of the irrigation plan for the Alentejo region, in 1957), the technicians considered that these measures were not enough. Mainly because, among other problems, the dictatorial governments hindered land restructuring and did not promote rural extension.

The fall of the regime in 1974 opened new possibilities to make more effective the modernisation model defended by the Centre, especially as the land reform was a part of the political programme of the various governments. Expectations that it would be possible to apply the agricultural modernisation model defended by the CEEA agronomists remained until the late 1970s, when changes in the agrarian reform and the agreements with the European Economic Community gave a different direction to the agricultural and rural policies. At the same time, both the services created directly by the state (for example, the rural extension service and the Instituto Nacional de Investigação Agrária), and the restructuring of the higher education system (new courses, new departments, new universities) took on the functions which competed with the activities performed by the CEEA. In the 1980s, the CEEA sought various ways to reposition itself in the new political and institutional framework, highlighting the high qualifications, experience and international recognition of its members. Some of these ways were the consolidation of advanced technical training (conferring PhD degrees, for example), the expansion of the think tank functions, by bringing together experts with the ability to promote research, consulting and conducting useful discussions to inform public policies.

At the end of the 1970s, the publication of several studies allowed a global view of the primary sector path and dissipated any optimism about agriculture performance (Caldas, 1978; Girão, 1980). The Agricultural Gross Product had stagnated and agriculture had not contributed to meet the growing food needs of the Portuguese inhabitants, requiring an increase in the import of basic products.

Concluding remarks

For nearly 30 years, the Centro de Estudos de Economia Agrária sponsored scientific research about Portuguese agriculture and rural society and raised relevant discussions about the development of the country. These activities did not have a direct impact in farm modernisation, but were instrumental in forming specialised elites and contributed to the expansion of rural studies.

During the dictatorship, the political meaning of the creation of the CEEA and the various interpretations afforded by the actions and speeches of those who were connected to the Centre, were not ignored in the arena of debate. On the one hand, the interventions of the various personalities linked to the Centre show that its existence was a constant factor of tension with members of the governing bodies. Researchers of CEEA had to find ways to avoid relationship conflicts with official bodies related to agriculture and even with members of the elite who occupied the most important positions in the organisations of the Nation. On the other hand, some of the proposals that emerged from the scientific research carried out at the Centre were used in the arguments produced by the opposition movements that fought against the Estado Novo regime. In some cases, personalities from the CEEA also developed civic actions in informal organisations that worked between legality and opposition to the regime.

The drivers of the Centre could be convinced, as was common at the time, that *good technique and progress* do not have necessarily a *political colour*. But in fact, the fulfillment of the purposes for which the Centre was created would lead to structural changes in the primary sector. And advocating measures that would lead to the implementation of this kind of changes had, even if implicitly, political purposes.

Following the dominant doctrines in Western Europe, the sponsors of the CEEA understood that the State should have an active role in driving the economy. And, as the entity that watched over the *common good*, would promote the most appropriate political decisions for each sector. The CEEA was endowed with logistical, human and financial conditions to produce scientific information that would allow the state to make the best decisions. In this sense, the Centre approached the think tank organisations that have emerged in other countries. Although developing other activities, the ambition to participate in the definition of public policies seems to have been the most constant objective of the CEEA.

Bibliography

Baptista, F. Oliveira. 2011. *Agriculturas e territórios.* Oeiras: Celta Editora.

Barreto, António, coord. 2007. *Fundação Calouste Gulbenkian: cinquenta anos 1956-2006.* 2 V. Lisboa: Fundação Calouste Gulbenkian.

Barros, Henrique de, *et al.*, 1983. *Centro de Estudos de Economia Agrária: 25 anos.* Oeiras: Fundação Calouste Gulbenkian.

Barros, Henrique de, *et al.* 1960. *Curso de gestão da empresa agrícola: sessões de abertura e de encerramento.* Lisboa: Fundação Calouste Gulbenkian.

Brito, J. Pais de, F. Oliveira Baptista, e Benjamim Pereira, coord. 1996. *O voo do arado.* Lisboa: Museu Nacional de Etnologia/Instituto Português de Museus/Ministério da Cultura.

Caldas, E. de Castro. 1978. *A agricultura portuguesa no limiar da reforma agrária.* Oeiras: Centro de Estudos de Economia Agrária/Instituto Gulbenkian de Ciência.

Caldas, E. de Castro, Mário Pereira. 1959. *Gestão da empresa agrícola: suas relações com a investigação económica e a vulgarização.* Lisboa: Fundação Calouste Gulbenkian.

Delicado, Ana, Raquel Rego, Cristina P. Conceição, Inês Pereira, e Luís Junqueira. 2012. *Ciência, profissão e sociedade. Associações científicas em Portugal.* Lisboa: Imprensa de Ciências Sociais.

Freire, Dulce. 2011. "Produzir mais e melhor: Estado, agricultura y consumo alimentario en Portugal (1926-1974)". *Ayer: revista de historia contemporânea.* V (83) 3.

—. 2008. "A campanha do trigo". *In Os anos de Salazar.* Coord. António do Paço. Lisboa: Planeta Agostini.

—. 2007. *Portugal e a terra. Itinerários de modernização da agricultura na segunda metade do século XX.* Dissertação de doutoramento dactilografada. Lisboa: Faculdade de Ciências Sociais e Humanas/Universidade Nova de Lisboa.

—. 2004."“Os baldios da discórdia: as comunidades locais e o Estado”. *In Mundo rural. Transformação e resistência na Península Ibérica (século XX)*. Coord. D. Freire, I. Fonseca, P. Godinho. Lisboa: Edições Colibri.

—. 2002. “Da ditadura à democracia. Continuidades e mudanças na política agrária durante o PREC”, *Arquivos da Memória*, nº12/13.

Freire, Dulce, Daniel Lanero. 2013. “State and agricultural modernisation in the Iberian dictatorships after Second Word War”. *In Integration through subordination. The politics of agricultural modernisation in industrial Europe*. Ed. Peter Moser and Tony Varley. Turnout: Brepols Publishers.

Fundação Calouste Gulbenkian. 2010. *Calouste Sarkis Gulbenkian. O homem e a sua obra*. Lisboa: Fundação Calouste Gulbenkian.

Galvão, Amândio. 1978. *Sobre a experiência portuguesa do conselho de gestão ao agricultor*. Oeiras: Centro de Estudos de Economia Agrária/Instituto Gulbenkian de Ciência.

Girão, José A. 1980. *Natureza do problema agrícola em Portugal (1950-1973): uma perspectiva*. Lisboa: Centro de Estudos de Economia Agrária/Instituto Gulbenkian de Ciência.

Glaeser, Bernhard, ed. 1987. *The green revolution revisited: critique and alternatives*. London: Allen&Unwin.

Griffin, Keith. 1979. *The political economy of agrarian change: an essay on the green revolution*. 2ªed. London: Macmillan.

Hewins, Ralph. 2009. *Calouste Gulbenkian 1869-1955. O senhor cinco por cento*. Lisboa: Texto.

Lains, Pedro. 2009. “Agriculture and economic development in Portugal, 1870-1973”. *In Agriculture and economic development in Europe since 1870*. Ed. P. Lains and Vicente Pinilla. London: Routledge.

Lanero, Daniel, Dulce Freire. 2011a. “Modernizar sin democratizar: sobre la industrializacion de la agricultura en las dictaturas ibéricas (1946-1975). Introducion”. *In Agriculturas e innovacion tecnologica en la Peninsula Iberica (1946-1975)*. Coord. Daniel Lanero and Dulce Freire. Madrid: Ministério de Agricultura, Alimentación y Médio Ambiente.

Lanero, Daniel, Dulce Freire, coord. 2011b. *Agriculturas e innovacion tecnologica en la Peninsula Iberica (1946-1975)*. Madrid: Ministério de Agricultura, Alimentación y Médio Ambiente.

Rollo, Fernanda. 2007. *Portugal e a reconstrução económica do pós-guerra: o Plano Marshall e a economia portuguesa dos anos 50*. Lisboa: Ministério dos Negócios Estrangeiros.

Scott, James C. 1998. *Seeing like a State: how certain schemes to improve the human conditions have failed.* New Haven: Yale University Press.
Stone, Diane, Andrew Denham, ed. 2004. *Think tank traditions. Policy research and the politics of ideas*. Manchester: Manchester University Press.

PART II

CONTRIBUTIONS OF SCIENTIFIC ASSOCIATIONS TO PROFESSION AND SOCIETY

CHAPTER SEVEN

SCIENTIFIC ASSOCIATIONS: A SOCIAL MOVEMENT IN EXPANSION BUT WITH LOW AMBITIONS?

RAQUEL REGO

Introduction

According to various different sources, Portugal ranks among low civic participation countries, often measured through recourse to an indicator on citizen association membership. In addition to the important qualitative differences these sources commonly overlook, such as the minimum number of individuals needed to establish an association in each country or the non-evaluated informal participation, these sources face a further reliability problem. In fact, there are some crucial limitations to data collected on voluntary associations in Portugal, like data dispersion or the lack of updated information (Rego 2010). Moreover, there are some associations that are not attributed any particular status, rendering them "invisible" to research, and that is the case with Portuguese scientific associations.

Initial research on this social movement was undertaken by the SOCSCI[1] project between 2010 and 2012 and represents one of the few sociological studies on Portuguese scientific associations generating quantitative outputs (Delicado *et al.* 2013). We have based this chapter on a section of the SOCSCI data.

Similar to other European countries, Portugal saw the emergence of learned societies in the early stages of the nineteenth century. These organisations corresponded to a far more closed type of association,

---

[1] SOCSCI was funded by the Portuguese Science and Technology Foundation (reference: PTDC/CS-ECS/101592/2008). For further information about the project, see its website: http://www.socsci.ics.ul.pt/english.

closely focused on scientific knowledge and with an expert membership. Today, we not only have many more associations, comprising far more diversified organisations while still focusing on scientific knowledge, but they also differ in their membership base, including researchers, laypeople, as well as science related occupational categories, such as grant holders. Another important observation concerning the scientific association movement is its increasing proliferation. In Portugal, associations are renowned for the extent of their fragmentation, and in some cases, such as trade unions and private professional associations, we even find parallel organisations, namely associations with the same goals and social composition but diverging ideological positions. The same also occurs with scientific associations, even if only in a minority of cases. Exit is always an alternative at voluntary associations (Hirschman 1970) and people therefore seek out associations reflecting their own points of view or alternatively set up a new subject branch within another association instead of nurturing factions or groups inside the original association. Opening up this kind of division inside associations is usually seen as an act of disloyalty in the social context in this southern European country. This also occurs when there are opposing election candidates whereas, on the contrary, this might be considered a core indicator of democracy (Lipset *et al.* 1956).

In this chapter, we analyse the somehow expected low level of this associative movement's professionalisation, as it remains young and still actively flourishing, while highlighting its causes and consequences. Furthermore, we demonstrate that there is no apparent ambition to become better institutionally rooted and to gain more influence in the public sphere.

We present a general quantitative overview of the professionalisation process, by using survey results addressing all scientific associations identified through to 2011 (n=301); in conjunction with another survey of a diverse sample of scientists (n=672); and complemented by qualitative data, collected by interviewing the leaders of a selection of 24 scientific associations heading a diverse range of organisations (in terms of subject fields and types, among other criteria).

A Recent and Heterogeneous Movement

Although with longstanding roots, with the first Portuguese scientific association dating back to the eighteenth century,[2] this social movement

[2] For further development of the historical approach, see Part I of this book.

experienced significant expansion in the 1970s in the wake of the establishment of the democratic regime in Portugal. In Portugal, legislation stipulating freedom of association arrived about seven months after the end of the almost fifty-year long dictatorship, in November 1974. The high number of scientific associations is consistent with the multiplication of associations in Portugal in that period. This trend also spread through other areas, precisely because one of the very first civil rights, the right to freedom of association only became practicable in this particular southern European country at that time.

Unlike other areas covered by a specific legal framework in Portugal, the country's scientific associations remain legally framed only within the general freedom of association legislation. Therefore, the SOCSCI project is responsible for the very first compilation of these associations into a database with no pre-existing means of accurately determining the number of scientific associations active in Portugal.

This does not mean that we cannot find any of these scientific associations included in the different and specific kinds of legal frameworks. In fact, some may simultaneously be considered not only scientific associations but also environmental associations, for instance. In this case, associations are endowed with a special status by the Ministry of Environment. Even though Portugal contains more than a dozen specific legal frameworks attributing special status with such recognition resulting in the right to consultation by public committee, the opportunity to benefit from public funding, etcetera, scientific associations do not show any intention of gaining a specific status.

The volume of scientific associations, nonetheless, seems far from negligible. The SOCSCI project identified over three hundred and sixty scientific associations in Portugal in a census carried out between 2011 and 2012. Despite the gap of about 10 years of difference, this number may prove surprising when compared with the United Kingdom (UK) which, according to the Scholarly Societies project,[3] had 316 associations in 2003. In fact one could expect a bigger difference when considering the population differential with the UK, home to a population around six times greater than Portugal's and with the former's longer and more prestigious scientific history.

Nevertheless, more associations do not necessarily mean a more active or influential movement. In fact, the end result might prove entirely the opposite. In Portugal, the sheer volume of these associations would seem to signal a not especially cohesive group. Inclusively, the trend towards

[3] Online at: http://www.lib.uwaterloo.ca/society/

the proliferation of Portuguese scientific associations may reflect a broader trait in Portuguese associative behaviours since we encounter the same pattern in other areas, such as occupational associations (Freire 2004) and trade unions (Alves 2009). In fact, in these two professional fields we may also identify several hundred associations with few organisational structures aggregating them.

This broad but shallow organisational finding may also be related to the early phase in social movement development and its still low level of ambition as regards influencing national policies for instance. In fact, in the beginning, organisations tend to focus more on attaining goals than cooperating with peer entities in order to gain a greater collective role.

The next graph sets out the growth in the number of Portuguese scientific associations over more than a century. The change triggered by the implementation of the democratic regime in the mid-1970s is quite clearly conveyed. Nevertheless, although some associations date back to the 1970s, the great boom in growth occurs from the 1990s on. In fact, the majority of scientific associations (63%)[4] was founded after that date, stemming from the development of the fields of science and scientific culture then taking place in Portugal (Gonçalves and Castro 2002). Also, the founding of associations of scientific professionals (of grant holders, researchers) has clearly benefitted from the increase in the system's critical mass.

Fig. 7-1 Accumulated number of Portuguese scientific associations by year of foundation (1901-2011)

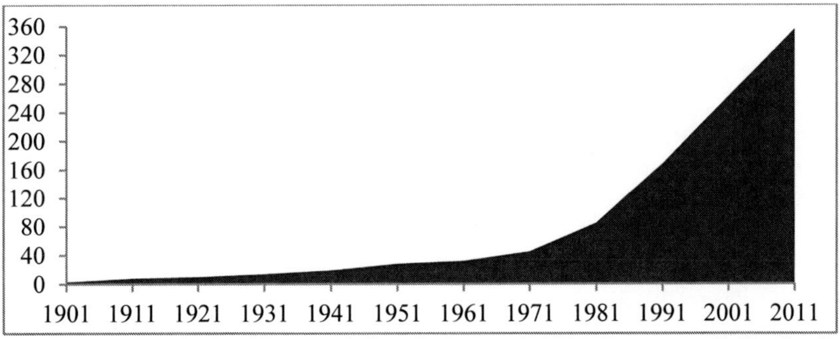

Source: Census of scientific associations in Portugal, 2011. Data unavailable for 60 associations.

[4] We decided to present integer numbers to enhance readability.

We would note that Figure 7-1 portrays the accumulated number of associations, since no data is available on their closure or inactivity. The SOCSCI project combined different sources, official and academic, in order to reach this distribution and we must also say that the concept of scientific associations applied in this research, despite some initial self-limitations due to the need to structure the study, covers a broad spectrum.

The amplitude of the scientific association concept adopted resulted from reflection on the need to incorporate the recent changes in the science field. We felt the need to reach beyond conventional analysis in which scientific associations are associations closed to only scientists or other learned people. In fact, the authors realised that, in addition to the two main types already identified by Evan Schofer (2003), a type of association needed to be more clearly distinguished, the professional associations.

In a publication on international scientific associations, Schofer (2003) distinguishes between two types of organisations:

a) "Professional scientific societies", which first appeared in the nineteenth century and are organised by disciplinary field; members are mostly actively engaged in science and generally not advising national governments;
b) "Socially oriented associations", associations founded in the wake of World War II, expanding rapidly, focused on the promotion of science as a means of combating social problems such as environmental degradation, and counting not only scientists but also on other citizens among their members.

The SOCSCI project also took these two types into account even while attributing more specific meanings. Based on our findings we also proposed what we believe to be more assertive expressions, for example "scientific culture" instead of "socially oriented". In fact, the first association type was named "disciplinary scientific societies" and corresponds to the oldest type of organisation, an association structured around a particular disciplinary field and with the connotation of a closed group as the word "society" seems to confer. The word "professional" was suppressed in accordance with our decision to establish a new category taking into consideration the Portuguese pattern of professional regulation and the rising relevance of science related occupations, to which we will return below.

The second type comprises "science dissemination associations" (see Chapter Nine in this book), more closely related with scientific knowledge than with issues stemming from solving social problems, as Schofer had proposed through the "socially oriented associations' category. The "science dissemination associations" group includes entities as diverse as astronomy clubs, organisations for the study and conservation of nature, archaeological associations or groups focused on disseminating technologies. Many associations falling into this category have been founded in response to initiatives launched under the auspices of the public agency Ciência Viva. This agency was set up in 1996 to foster citizen participation in science initiatives and raise public awareness about science-based knowledge. Its most visible campaign is arguably the "Science in Summer" programme (Conceição 2011). These campaigns usually required the existence of a non-profit collective entity in order to apply for funding. Consequently, many science dissemination associations were founded, demonstrating their inherently instrumental nature deriving from a top-down process.

By late 2012, the SOCSCI project had gathered a total of 362 scientific associations and the database is updated on a regular basis. Disciplinary societies proved the most numerous, with 263 organisations identified, while science dissemination associations accounted for a further 81 associations. However, we also identified associations focused on the defence of the interests of both scientists and other science-related occupations. Considering this research was undertaken in a country with a Latin language, the distinction between occupation and profession, the liberal professions in the Anglo-Saxon tradition of the sociology of professions, we considered a wider meaning for the therefore classified as a new organisational type: "associations of scientific professionals".

We found this new association type bore similarities with the two other types, in terms of undertaking vocational training, even though their broader concerns more commonly stem from working conditions or deontological practices. The "associations of scientific professionals" make up a small group of 18 associations, representing 5% of the total of scientific associations. The common category denominator is the prevalence of scientific research occupational groups, the majority of such associations (ten cases) are interdisciplinary in character. Moreover, the legal framework includes diverse types, thus not only private associations, but also public associations and trade unions.

The Emergence of New Occupations in Portuguese Science

The "associations of scientific professionals" were mostly founded, as with the majority of scientific associations in Portugal, in the post-1974 period within the framework of the country's newly founded democratic regime. We include here secondary school teachers' and university lecturers' associations, along with researcher, research assistant, grant holder, and professional associations in the narrow sense. The emergence of these later professional categories, which can be seen as a sign of the recent development of our scientific system, deserves in our point of view a special attention.

This narrow sense means we did include some, but not all, professional associations with self-regulation power. These associations do not represent always professionals in the narrow sense of the word, or to be more precise, in the Anglo-Saxon notion. Here, we must bear in mind that Portugal belongs to a pattern of professional regulation differing to the Anglo-Saxon norm. As Moran and Wood (2003) say, regulation is a political issue. In fact, self-regulating professional associations in Portugal, against what happens in the UK for instance, accumulate the regulation function with the representation one. Moreover, it may prove surprising to other European countries that some occupations are subject to regulation in Portugal while they are not in other countries, as is the case with economists. Economists may be considered to have a low public interest in order to obtain the self-regulating power from the state, however they got it in Portugal long time before psychologists or nutritionists, which are groups that deal with human health directly.

We sustain that the relative youthfulness of the scientific associative movement in Portugal, first of all, correlates with the expansion in the scale of science-based human resources, furthermore demonstrated by the emergence of associations centred on scientific professional occupational categories. Several such occupational groups have taken to organising associations whose vocation straddles the scientific association, professional association and trade union organisational structure. The "associations of scientific professionals" type shows in fact a hybrid nature.

This type is particularly characterised by three associations founded in the last decade and undertaking the interdisciplinary representation of specific groups (see next chapter for more details):

- the Association of Research Grant Holders (ABIC, Portuguese acronym hereafter) in 2003, particularly concerned with the precarious situations faced by grant holders, who have scarce social protection rights,;

- the Portuguese Association of Female Scientists (AMONET), which was launched in 2004 when the university system was already highly feminised; and
- the National Association of Science and Technology Research Fellows (ANICT) in 2010, established for the recently hired PhD researchers, on fixed terms contracts, under the governmental Commitment to Science Programme. In 2005, the Commitment to Science awarded research units the opportunity to hire full time researchers. Over 1,000 researchers were hired for five years between 2007 and 2009, enabling both the founding and expansion of scientific teams and scientists' mobility.

The SOCSCI database recently also incorporated a "science manager" association. Science managers represent a new occupational category that clearly reflects the significance of the investment made in the scientific system in recent years in Portugal as well as the consequent professionalisation of the field. Given the lack of quantitative data on this association during the fieldwork period, our findings do not yet include this association.

This national trend is in line with development in many other European countries. In fact, the creation of the European Council of Doctoral Candidates and Junior Researchers (EURODOC) in 2002 (Urani *et al.* 2004) is symptomatic of a similar social movement at the European level even if substantially lagging behind events in the USA.[5] EURODOC assembles young researchers from 35 European countries, including the EU27, except for Ireland. The organisation's purpose involves representing doctoral candidate and young researcher interests at the European level. It focuses on matters including education, career and professional development in addition to fostering the circulation of information on issues of relevance to young researchers and nurturing and establishing cooperation between national member associations.

An Associative Movement in the First Stage of Its Life Cycle

Although the SOCSCI project did identify an important number of associations, the majority has not yet overcome the first stage of organisational development. In fact, this becomes our conclusion after

[5] For details on the American experience, see Chapter Eight.

analysing the stunted nature of the movement taking into account what we name "the degree of the organisation professionalisation", hence, its bureaucratic level of complexity, degree of specialisation, and stage of institutionalisation (Rego 2010).

The clearest facet demonstrating the shallow nature of this social movement is its low level of professionalisation. While the majority of Portuguese scientific associations does employ paid workers, they employ neither significant numbers nor do they have many technical functions, as set out in the next table. On a similar note, we identified only one full time, paid manager among our 24 case studies, the professionalisation of the board level representing the highest indicator of professionalisation, as detailed elsewhere (Rego 2010).

Table 7-1: Paid and voluntary workers in Portuguese scientific associations in 2011 (%)

	Paid workers	Voluntary workers
None	36.7	53.7
One worker	27.8	4.9
Two or more workers	35.6	41.5

Source: Survey of scientific associations (2011); n=90

Another indicator of this associative movement's level of underdevelopment stems from the lack of federations active at the national level. In the early 1990s, the first and only Portuguese Federation of Associations and Scientific Societies (FEPASC) was established to "…contribute not only to intensifying relationships within the scientific community but also to engage in actions of scientific dissemination among the population" (Gonçalves 1993, 146) and to bring about "…genuine representation of the scientific community through their representative organisations in consultative bodies such as the Higher Council for Science and Technology, among others" (Gonçalves 1996, 59). In fact, FEPASC still retains the responsibility of appointing two members to the National Council of Education and several scientific associations claim to be FEPASC members, though the Federation is in practice fairly inactive (Oliveira 2001): in the past decade there were no reported or known events and it is difficult if not impossible to contact its board.

An umbrella organisation would be particularly important for political purposes, since it enhances collective action. Nevertheless, according to our interviews with the selected associations, there are no important complaints about their scant influence over public policies. Our scientific association survey findings point in a similar direction and hence our observation that this associative movement apparently lacks the ambition for a more relevant and influential role.

The low level of organisational complexity is demonstrable both through the limited financial and human resources available to these associations and, at least in most cases, by their lack of impact and action in the public sphere. We observe that these are the two most commonly mentioned problems identified by our survey: 66% of Portuguese scientific associations declare not having enough financial resources, with 29% reporting no human resources. These aspects in practice inhibit these associations from playing a different role in society.

At the meso analytical level, we also identify a structural weakness in this movement related to, on the one hand, the low level of cooperation between associations[6] and the absence of a dynamic federation; and, on the other hand, a system fairly dependent on public funding when this has always represented a limited source of funding and today, in the context of recession and financial crisis, particularly fragile. Nevertheless, according to our survey, public funding comes in third place in the list of sources of financing, behind fees and the association's own services. In fact, we cannot overlook how membership recruitment, and the consequent increase in fee revenues, often takes place at scientific meetings and conferences themselves often funded by public entities.

However, there are also external variables that may contribute to explaining the underdevelopment of Portuguese scientific associations as organisations. One such factor may result from the increasing importance of European and inclusively international associations (Evetts 2001), most of them with longer track records than the Portuguese associations in promoting scientific meetings and publications. This proves especially true when even the social sciences are following the evaluation parameters established for the natural sciences and correspondingly valuing publications in journals with a ranked impact factor, for instance. We would add Portugal currently does not have many social science publications in these indexes.

[6] In this sense, whilst the mutual ignorance was obvious, there was the interest in finding out about each other and exchanging experiences demonstrated by the scientific associations attending the workshop held by the SOCSCI project.

A Limited Influence beyond the Scientific Community

As with other voluntary associations operating solely under the legal framework established by the Portuguese Freedom of Association Act of 1974, scientific associations experience a deficit of power and social recognition in society, and sometimes also in their respective scientific fields. Nevertheless, according to our interviewees, they do not seem to actively seek to enhance their levels of influence.

Scientific associations may seek to influence policies in their particular field of representation, as well as defending the interests of their activities and members. This sometimes happens with scientific professional associations in particular regarding hiring policies, recognition of qualifications, budgetary issues, and etcetera. The forms of pressure are not always obvious: ranging from parliamentary hearings through to personal relationships.

We would note that usage of the term lobby still raises some resistance in Portugal. The connotation of strangeness the word lobby creates was directly referred to in the interviews. Words like lobby or pressure group are associated with a capacity for sanction that trade unions deploy through strike action, for instance. Therefore the word influence seems more consensual. The reality encountered at the European level again seems far distant. We would recall that there is already a public database of voluntary civil society organisations structured and identified as potential lobbyists of the European Commission. This registration aims to foster transparency instead of disguising the existence of these organisations and groups.

The lobby concept does seem to be accepted as a tool for action in the case of some scientific professional associations. The association positions may take on a specific character, focusing on representing certain groups within the scientific community. For example, AMONET has played an important role in raising awareness among the scientific community about the difficulties experienced by female researchers in reaching senior positions in higher education teaching and research career structures. Its action remains fairly circumscribed to this gender issue and especially deploying collective actions based on a formal repertoire.

According to the survey of the scientific associations identified in the SOCSCI project, representation of researchers' interests has been an activity performed by slightly over half of all associations in recent years and meaning that even discipline based societies and science dissemination associations engage in this task. This may not be their central activity but it is also not a negligible facet when it comes to analysing the role of scientific associations in Portuguese society. However, as the next table shows, around half of Portuguese scientific associations produce only more occasional opinions than take up a permanent seat on any public committee.

Table 7-2: Political influence of scientific associations by association type in 2011 (%)

		Disciplinary scientific societies	Professional associations of scientists	Science dissemination associations	Total
Participation in consultative public committees	Regularly	7.3	27.8	12.5	12.4
	Occasionally	36.4	**44.4**	28.1	35.2
	Never	**56.4**	27.8	**59.4**	**52.4**
Opinions and counselling on public policies	Regularly	16.4	**50.0**	15.6	21.9
	Occasionally	**50.9**	44.4	**53.1**	**50.5**
	Never	32.7	5.6	31.3	27.6

Source: Survey of Portuguese scientific associations (2011); n = 105

In this sense, and to take an example, ANICT, which we would recall is a very recent association composed of Portuguese and foreign researchers working in Portugal, frequently contacts the public authorities to lobby in favour of the employment continuity of researchers hired under the Commitment to Science Programme. This association paid particular attention to the proposal of an evaluation system as a mechanism designed to bring about contract renewal for at least the best among these researchers. Very recently, in 2012, ANICT, for instance, has produced a document entitled a Plan for Excellence in Research, which contained a concrete set of proposals. Nevertheless, the continuity of researchers' employment has not been assured and the end of the contracts is coinciding with the crisis period.

On the other hand, the ABIC repertoire includes more visible forms of protest such as street demonstrations, petitions, press releases and press conferences. A clear illustration of this kind of collective action, more centred around mobilisation than negotiation, to quote Claus Offe (1985), and also more creative in order to the capture the media's attention, is the "Fotoprotesto" held in May 2011. This initiative consisted of group photographs of grant holders from various scientific institutions, holding up placards to draw attention to the social injustice of these scientific occupations as they neither have the right to full social protection nnor they make tax payment and remain in this precarious situation for numerous years despite its supposedly temporary nature. The photographs were widely publicised in the media and ABIC held a press conference in front of their funding entity, the governmental Foundation for Science and Technology, premises.

Meanwhile SNESup, a trade union and professional association representing faculty professors and researchers, seeks action focused especially on behalf of higher education lecturers, the most representative group among its membership. In this sense, the restructuring of the University and Polytechnic Teaching Career Statute has become one of its main fields of intervention. Collective SNESup action has tended to take a more institutionalised approach taking into account its legal framework stemming from its trade union status. As a trade union, it has the right to be consulted in certain advisory councils and may convene strike action as a means of exerting pressure. Furthermore, SNESup was the only association of the associations of scientific professionals that mentioned recourse to legal action. However, we would point out that such court action does not prove either efficient or effective in Portugal and requires resources usually unavailable to scientific associations.

Our case studies open up an insight into the successful application of pressure, with the main target obviously being the aforementioned Foundation for Science and Technology, the main executor of public funding for research and development policies in Portugal, although the Ministry of Science holds ultimate responsibility for the national scientific and technological system. Moreover, associations also target the ministries that hire scientific professionals, like the health and education ministries. Simultaneously, there are also cases where scientific associations unsuccessfully persist in their claims and demands, a factor likely to be on the increase given the current context of crisis and the consequent cuts to science funding.

Within the scope of opinions released on the associations' own initiatives, we find that their output is increasingly required by political decision makers within a period in which science is practically "sacralised". In several domains, scientific studies are already legally required to support political decision-making. This is the case of environmental field where an impact study is mandatory, or in the case of the professional field, where a study is needed to support the decision of members of the national parliament on the delegation of self-regulation power.

Future research should probably investigate if scientific associations are losing space on behalf of individuals, and supranational associations, especially in small countries like Portugal with less than forty years of democracy and a young scientific system. And verify if research units work better as a bridge taken to reach out to experts.

Final Remarks

The SOCSCI project made the first ever census of this field and ascertained that there is an important scope of organisations falling into the category of scientific association. We may also identify a clear bond between recent Portuguese history and the emergence of diverse scientific association types. They still are in a first stage of development, above all because they neither access significant levels of resources nor do they report effective levels of professionalisation.

The main differences between the three types of associations identified remain based on the respective social composition and its goals. Hence, we may differentiate between the old, elitist societies that retain a kind of closed-shop screening process–members enter into contact with these associations mostly through professors (38%)–, and the more recent associations that even accept non-scientists as members–in particular, the scientific dissemination associations and associations of scientific professionals emerging out of the bureaucratisation of science–and especially discovered through a colleague (34%) or a friend (23%), respectively.

Furthermore, closer analysis reveals how the field of science has expanded substantially since the 1990s, driving the emergence of new professional occupations. Therefore, we do not have just two main association types but rather three: scientific societies, scientific dissemination associations, and the associations of scientific professionals.

Potentially due to the increasing importance of the supranational scope of science and its professions, Portuguese associations seem to lack any ambition to grow in recognition and influence, even though an important percentage of our respondents to a survey of scientists (40%) maintain that they either do not or do not entirely agree that national associations are losing in importance due to the internationalisation of science.

Another factor we consider helps explain the low profile of scientific associations in Portugal is how the scientific field today ranks alongside other actors of equal or higher importance such as the research units to which experts are more closely related. The scientific career trajectory is not as clearly defined by membership of associations as by the ties to research units or universities. However, some apparently still do play a "trampoline" role for further career developments and at least generating important opportunities for networking.

Portuguese scientific associations are deeply intertwined with the national scientific field. They play a major role in circulating knowledge among peers and the socialisation of younger practitioners. Paradoxically, or perhaps not, they appear to develop stronger links with international associations than with their national counterparts. In fact, this seems to be a proliferating but non-cohesive movement without any indication of mutual organisational cooperation in order to be heard by public policies.

Bibliography

Alves, P. M. 2009. "*Cidadãos e Militantes: Uma Contribuição para as Teorias da Acção Colectiva Sindical.*" PhD dissertation. ISCTE-IUL.

Conceição, C. P. 2011. "*Promoção de Cultura Científica. Análise Teórica e Estudo de Caso do Programa Ciência Viva*". PhD dissertation. ISCTE-IUL.

Delicado, A., R. Rego, C. P. Conceição, I. Pereira & L. Junqueira. 2013. *Ciência, profissão e sociedade: associações científicas em Portugal*, Lisbon: ICS.

Evetts, J. 2001. "Professions in European and UK markets: the European professional federation". *International Journal of Sociology and Social Policy* 20:11/12, 1-30.

Freire, J. (ed.). 2004. *Associações profissionais em Portugal*, Oeiras: Celta.

Gonçalves, M. E. 1993. "Ciência, comunidade científica e democracia em Portugal". In M. E. Gonçalves (ed.). *Comunidade Científica e Poder*, Lisbon: FPASC/Edições 70, 133-150.

—. 1996. "Mitos e realidades da política científica portuguesa." *Revista Crítica de Ciências Sociais*, 46, 47-67.

Gonçalves, M. E. & P. Castro. 2002. "Science, culture and policy in Portugal: a triangle of changing relationships?" *Portuguese Journal of Social Sciences*, 1(3), 157-173.

Hirschman, A. O. 1970. *Exit, voice, and loyalty: responses to decline in firms, organisations, and states.* Cambridge, MA: Harvard University Press.

Lipset, S., Trow, M. & Coleman, J. S. 1956. *Union democracy: the internal politics of the international typographical union*, New York: New York Free Press.

Moran, Michael & Wood, Bruce. 2003. *States, regulation and the medical profession*, Buckingham–Philadelphia: Open University Press.

Offe, C.1985. *Disorganized capitalism*, MIT Press.

Rego, R. 2010. *Dirigeants associatifs: Engagement et professionalisation*, Paris: l' Harmattan.

Schofer, E. 2003. "International Science Associations, 1870-1990". In Drori, G. S. *et al.* (eds.). *Science in the Modern World Polity: Institutionalization and Globalization*, Stanford CA: Stanford University Press, 81-99.

CHAPTER EIGHT

PARTICIPATION OF RESEARCHERS IN PROFESSIONAL ASSOCIATIONS

LUÍS JUNQUEIRA

Introduction

The growth of the Portuguese scientific system over the past decades is undeniable, reaching a volume of dedicated human resources that would seem unthinkable a few decades ago. But this growth spurt was also responsible for highlighting some negative trends in the recent development of scientific organisation in post-industrial societies, mainly the consequences on the careers of young researchers who find increasingly difficult to achieve a stable job in academia. This development led to the formation of two new organisations for the purpose of representing and defending the socioprofessional interests of these new researchers, the Association of Research Grant Holders (ABIC), established in 2003, and the National Association of Science and Technology Researchers (ANICT), established in early 2010 (see also the previous Chapter in this book). As such, it is important to give an account of the development of this new associative movement within the scientific community. This chapter aims to contribute to the understanding of this new social phenomenon by presenting an analysis of these two new associations of Portuguese scientists, through their membership composition and participation of their members.

The first part of this chapter will briefly describe the emergence of these new associations, against the background of the Portuguese scientific system. The analysis of the composition and participation of the members of these associations will follow in the second part.

This paper is based on a research developed as part of the MSc programme in Sociology at the Faculty of Social Sciences and Humanities, New University of Lisbon and is part of the research project SOCSCI

Scientific Societies in Contemporary Science (PTDC/CS-ECS/101592/2008), funded by the Foundation for Science and Technology and carried out at the Institute of Social Sciences, University of Lisbon.

Framework

The emergence of protest movements formed by the scientific community is not an isolated phenomenon in the professional world. The impact of political mobilisation during the 1960s made collective action more attractive to certain social groups as a form of political participation and lobbying (Meyer and Staggenborg 2011). It is important to account for the change of protest against economic power (industry) to protest against political power (government). In particular, in protest movements associated with labour, the main difference from the classical labour movement is the change of focus from the class opposition between labour and capital to a focus on the policy-making processes on which certain professional groups, such as teachers or civil workers, are dependant (Porta and Diani 2006). The labour rights movement—and the associations that represent them—,despite a decline in the 1980s, have shown some signs of revival since the late 1990s (Clawson and Clawson 1999; Frege and Kelly 2003).

There is already a significant body of literature concerning the unions of graduate employees that focuses mainly on the formation of these new organisations. These new unions seek to gather together PhD students, especially those who are integrated in universities as teaching assistants or research assistants (Rhoads and Rhoades 2005). The first American graduate union was created and recognised in 1969 at the University of Wisconsin-Madison, due to a favourable stance of the university administration. The influence of student protest in the late 1960s and early 1970s will bring nine more unions until 1975. But the graduate union movement would slow down during the next 15 years and would only make a return in the 1990s. In the period between 1990 and 2002, 16 new unions were recognised by their universities, more than the total of existing unions at the beginning of that period (Singh, Zinni, and MacLennan 2006) and this process showed no signs of slowing down in 2003, with several campaigns for the formation of new unions under way during that year (Kerchner 2003). The movement grew not only in the number of organisations but also in the number of graduate employees covered by unions, which rose from 14,000 in 1990 to nearly 40,000 in 2000. Another important event was the recognition of the first union at a private university in the USA—the New York University (NYU) in 2000.

The formation of this union involved the presentation of the case to the National Labour Relations Board, responsible for regulating labour relations, which forced NYU to recognise the right of these workers to collective bargaining with the university administration. This process eventually encouraged more disputes over the formation of unions in other private universities (Rhoades and Rhoads 2002).

Among other reasons, these associations were a response to the changes that occurred in the American university, namely the growing number of researchers and PhD programmes and post-doctoral level, corresponding to the extension of a precarious employment situation dictated by the lack of entry-level positions in the teaching career (Julius and Gumport 2002; Hoeller 2007). Several studies characterise these changes as particularly adverse to the working conditions of doctoral students and postdoctoral researchers. Hoeller (2007) classifies this new organisation of academia as a two-tier system, in which a small number of professors with academic tenure are sharing space with a growing number of young researchers and teaching assistants with temporary labour ties and limited opportunity to access the positions that would allow them to progress in their careers. In short, a new group of researchers has emerged in academia, one who shares a set of professional interests that diverge (or clash in some cases) with those of pre-existing groups. The fact that existing organisations, such as faculty unions, were not suitable to represent the interests of these young researchers led to the creation of these new organisations (Kerchner 2003).

Portuguese science had an impressive growth spurt at the beginning of the century, heavily influenced by the EU's Lisbon Strategy, a set of societal objectives aimed at strengthening the position of Europe as a knowledge-based economy. It included the ambitious goal of increasing European R&D spending to 3% of the gross domestic product by 2010, far from the slight increase of between 1.9% and 2% that actually occurred. However, in the Portuguese context, the situation was quite different. During the 2000s the growth of the national scientific and technological system has not only continued but also experienced a sharp increase, especially since 2005 with the start of the "Commitment to Science" Programme, which reinforced the centrality of science and innovation in the socioeconomic development strategy of the country. R&D investment, which had already had a period of significant growth during the 1990s, thus converged with the EU average, doubling from 0.8% in 2001 to 1.6% of GDP in 2010. The growing number of researchers in Portugal accompanied this growth, from just under 17,000 to about 45,000 in 10 years. The growth in financial and human resources was translated into an increase in scientific output of universities and research centres in Portugal. The number of articles published in international scientific journals continued to show a growing trend, quadrupling to a total of 13,424 in 2010. Also significant is the growth in the number of patents granted to Portuguese universities and research centres, from being almost negligible at the beginning of the decade to about 70 per year in the past four years.

The formation of new professional associations also benefited from the existence of previous organisations, mostly based on an established class of higher education teachers and full time researchers in public national laboratories. Two of these organisations are a product of the initial organisation of the Portuguese union movement during the period following the establishment of the democratic regime in 1974. The first was the National Teacher's Federation (FENPROF) formed in 1983, as a federation of several regional teachers' unions, which was created to encompass all degrees of education, including higher education. FENPROF maintains a dedicated section in its organisational structure for higher education and research issues. The other one is the Scientific Workers' Organisation (OTC) whose creation is marked by the existence of two separate careers related to research on the national science and technology system, the higher education lecturer and the scientific researcher. Full-time researchers working in the various state laboratories, who were in much lower number than the teachers in higher education institutions, had more difficulty in seeing their interests represented by the teachers' unions emerging in this period. The association was created in 1979, and despite its objectives of professional representation, never acquired official union status. A third important organisation is the National Higher Education Union (SNESup), which was formed in 1989 as a secession of the teachers' union movement and with the aim of creating an organisation fully dedicated to the representation of higher education teachers. Today, SNESup is the largest union in the higher education area, with more than 5,000 members.

Methodology

The empirical evidence for this chapter is based on an online survey of ABIC and ANICT members. The associations, at the request of the research team, used their existing mailing lists to disseminate the survey among their members and a survey platform was used to collect the answers (SurveyMonkey). The survey of ANICT members was conducted between February 25 and March 9, 2011; 73 valid responses were received, corresponding to a response rate of 22.7%. The survey of ABIC members was conducted between May 19 and July 13, 2011; 125 valid responses were obtained, corresponding to a response rate of 17.9%.

It is difficult to assess the quality of the samples. To begin with, one must consider that these samples were not collected by a random method and therefore the inference of their representativeness from probabilistic methods is excluded. The response rate to both questionnaires can be considered in line with what is usual for this type of survey. The literature conducted to evaluate this method found response rates ranging between 19.3% and 47.2% in email questionnaires in which the respondent base consists of students of higher education or teaching and research staff in universities (Kaplowitz 2004; Sheehan and Grubs 1999). Despite these numbers being higher than those presented in this work, it must be borne in mind that response rates to email surveys have been declining since the mid-1990s. Despite the fact that greater use of email should increase response rates, certain factors such as the upsurge in average daily emails received by users, the trivialisation of spam (unsolicited mail), or the loss of a certain effect of novelty have contributed to a decline in response rates (Sax, Gilmartin and Bryant 2003; Sheehan 2006).

It should also be taken into account the possibility of bias arising from factors other than the propensity of certain groups of individuals to respond to this type of questionnaire. First, certain factors generally considered are not necessarily problematic in this situation. The level of education is expected to be high among the target population (Sax, Gilmartin, and Bryant 2003), since they are members of professional scientific associations, and the access and ability to use email (Zhang 2000) should be widespread among them. Another aspect that is also identified as relevant is the "prominence" of the investigation among the target population, that is to say, the interest raised by the research subject research (Sheehan, 2006). In this case there may be a bias resulting from greater disposition of those members more interested or involved in the associations to respond to the questionnaire. The age of respondents is also an important aspect, because there is a higher propensity of young adults to respond to questionnaires by email, compared to older individuals (Sax, Gilmartin, and Bryant 2003).

In-depth interviews with the presidents of the governing boards of both associations were also conducted, in order to collect information about the activities and internal organisation. This research also accessed the documentation available at the associations' websites, such as communications, proposals and letters of demand and public documentation on science policy.

The new researchers' associations in Portugal

The formation of ABIC (Association of Research Grant Holders) follows the rapid growth of Portuguese science that occurred during the 1990s. This period witnessed a dramatic increase in the number of doctoral and post-doctoral grants awarded by the Foundation for Science and Technology (FCT) (the main public science funding institution) and the strengthening of the funding for research projects and R&D institutions. It is also in this period that an important step in the regulation of post-graduate training is taken, with the approval in 1999 of the first statute of the scientific grant holder, in which some labour rights of the grant holders are recognised (Decree 123/99).

The formal recognition of the grant holder status was quickly transformed into a political resource by this "new" category of scientific workers. By late 2001, the Platform of Research Grant Holders was formed under the initiative of a group from the Faculty of Sciences of the University of Lisbon, who prepared a letter of demand at the beginning of the following year, focused on contesting the newly approved statutes.

During the year 2002, the platform held meetings in several higher education and R&D institutions in the Lisbon area, in which the letter of demands was presented. These meetings would culminate in a national meeting of grant holders in February 2003, during which the guidelines for the Association of Scientific Grant Holders (ABIC) were approved.[1]

The formation of the National Association of Science and Technology Researchers (ANICT) is closely connected to the "Commitment to Science Program", a set of initiatives by the Ministry of Science aimed at bolstering human resources in science, launched in 2007. Among these measures was the creation of competitive calls allowing the hiring of full-time Research Fellows by R&D institutions, with the objective of "increasing critical mass or creating new teams, as well as the mobility of researchers".[2] The costs with these new contracts would be supported by the Foundation for Science and Technology, complemented by a variable percentage of contribution paid by the host institutions.[3] Two calls were issued, leading to the recruitment in 2007 and 2008 of 1,185 researchers, on five-year contracts.

---

[1] http://www.bolseiros.org/pdfs/mocao.pdf

[2] MCTES, A commitment to science for Portugal's future

[3] Regulation Ciência 2007 http://www.fct.pt/apoios/contratacaodoutorados/regulamento2007.phtml; Ciência 2008 http://www.fct.pt/apoios/contratacaodoutorados/regulamento.phtml

ANICT was created with the purpose of providing political representation for these contract researchers, who find themselves in a different situation from that of tenured researchers and professors. The process that will lead to the founding of ANICT starts at Minho University, in March 2009, when a group of contract researchers hired under the "Commitment to Science Programme" met to form the Network of Research Fellows of the University of Minho. In May, the group organised a national meeting of the "Commitment to Science" researchers, in which about 100 researchers from around the country came together to discuss their professional situation.[4] This meeting also had the participation of members of the parliamentary committee on science and education.

Another important initiative of the ANICT founding group was to conduct a survey of the working conditions of these researchers in 2009. The survey ascertained the difficulties of integrating these researchers in the host institutions, which, it appears, in some cases were not prepared to receive them: many reported insufficient working and research conditions, such as lack of offices, laboratory space, equipment, materials, reagents, etc. Another aspect that stands out is the uncertainty of the researchers regarding their professional future. Although 81% of respondents to the ANICT survey expressed a desire to remain in the host institution, 65% said they were uncertain of the future of their professional situation after the end of the five-year contract. This sense of professional uncertainty would lead this group to officially form ANICT in January 2010.

Participation in the associations

Both associations are mainly composed of researchers working at universities and R&D institutions, where human resources in research have grown over the past 20 years (Table 8-1). A national policy for science and technology contributed to consolidating the central role of universities in scientific production, which over the past 20 years has remained as the sector with the greatest impact on the growth of human resources in science.

The successive advanced training programmes have mainly favoured the expansion of research centres and institutions of higher education. This predominance of academia is therefore reflected in the formation of these associations. As expected, members of the two associations roughly correspond to the groups from which each association stemmed: the grant

[4] ANICT Symposium 2010 Booklet

holders in the case of ABIC and contract researchers in the case of ANICT.

Table 8-1: Professional situation of members of the association (%)

ANICT		ABIC	
Researcher in R&D or higher education institution	83.3	Grant holder in R&D or higher education institution	78.3
Researcher in another kind of organisations	9.1	Researcher in R&D or higher education institution	9.4
Grant holder in R&D or higher education institution	6,1	Grant holder in private non-profit institution	6.6
Other	1.5	Other	5.7
N	66		106

It is interesting to note that the associations focus mainly on the professional groups that presided over their creation, particularly in the case of ANICT that seeks to represent all non-tenured researchers with a PhD, including postdoctoral fellows. In its statutes and regulations, ABIC reinforces the idea that it is mainly an organisation for representing grant holders, but ANICT, at least officially, presents itself as an organisation for all researchers with a PhD degree.

> Most people who are involved... the vast majority of members are researchers with fixed-term contracts, is it not? So, people who recently became researchers. (…) from the [Commitment to] Science Programme and before, from the Associated Laboratories (…) it was one of the means of hiring before the Science Programme. But we have some researchers with permanent contracts who are members of the Association. Not many who are actively involved in the management of the Association, in fact, but we have members who have permanent contracts. (ANICT Interview)

Despite this broad scope, the association seems to have a limited appeal for tenured researchers and post-doc grant holders, since its growth was mostly based on contract researchers who share the same professional issues as the organisation's founding group. In this case, ANICT overlaps with pre-existing organisations, such as the National Higher Education

Union (SNESup) and the aforementioned Association of Research Grant Holders (ABIC).

Both associations show a distinctive and fairly condensed age profile (Table 8-2), reinforcing the idea that they represent researchers at particular stages of their careers. Most of the respondent members of ABIC are aged between 26 and 35 years, which roughly corresponds to the period during which young researchers participate in PhD training programmes and post-doc fellowships supported by research grants. The age profile of ANICT members shows, in turn, a group roughly 10 years older and therefore at a later stage of the research career. The profile of the founders of ANICT shows that most of them had fixed-term contracts or postdoc grants before applying for their current position.[5]

Table 8-2: Age of members by association (%)

	ABIC	ANICT
Less than 25 years	0	2.6
26-35 years	28.8	70.1
36-45 years	62.1	24.8
46-55 years	7.6	2.6
More than 55	1.5	0
N	66	106

The distribution of respondents by region of residence largely reflects the centrality of some of these areas in the Portuguese scientific system (Table 8-3). The region of Lisbon and Tagus Valley, which is undoubtedly the largest centre of scientific production, with three public universities, one of them the largest in the country, over the past 15 years has received between 40% and 50% of the funding for R&D institutions, projects and grants. The North, meanwhile, comes in second place in the distribution of R&D resources, with about half the funding received by Lisbon. Members residing in the areas of Lisbon and Porto, which concentrate most of the human resources in research, represent over half of the total in both associations. The associations also seem to maintain a profile marked by their geographical birthplace. In ABIC, created in Lisbon, the members from Lisbon have a very significant prevalence over the total of members. Conversely, ANICT, founded mainly by researchers from the northern region of the country, maintains a higher proportion of members from those areas, especially Porto and Minho.

[5] ANICT Symposium 2010 Booklet

Table 8-3: Area of residence of members by association (%)

	ANICT	ABIC
Lisbon Metropolitan Area	34,4	45,0
Porto Metropolitan Area	26,6	11,7
Minho	9,4	5,4
Coimbra	7,8	8,1
Algarve	4,7	5,4
Aveiro	4,7	5,4
Other areas	12,6	18,9
N	64	111

The distribution of surveyed members by subject area of training (Table 8-4) shows a tendency to attract researchers from particular scientific areas. The natural sciences are by far dominant among ABIC members and there are fewer members from engineering and technology and the health sciences. The profile of ANICT members, albeit with different characteristics, shows also a dominance of the natural sciences, while the humanities and social sciences are less represented.

Table 8-4: Scientific area of last completed degree of members by association (%)

	ANICT	ABIC
Natural Sciences	65.0	56.7
Health and Medical Sciences	5.0	2.7
Engineering and Technology	11.7	8.1
Social Sciences and Humanities	15.0	33.3
N	60	111

This disciplinary concentration in the natural sciences is interesting when confronted with the literature on unions of young researchers in North America. In particular, it is mostly among young researchers from the humanities and social sciences that we find the support for American teaching assistant unions. The reasons for this phenomenon relate to: a) the greater difficulty for the students of social sciences and humanities in finding employment outside the university b) the major obstacles that senior researchers in the social sciences and humanities face when trying to obtain funding for research by comparison with their counterparts in the more "instrumental" sciences (with greater technology integration and expectation of economic return), a barrier that threatens the continuity of research work after completion of the PhD (Ehrenberg *et al.* 2004), c) lower stipends for research grants in the social sciences and humanities (Lafer 2003), and d) a higher workload of teaching activities for the social sciences and humanities, resulting in longer completion times for PhD degrees in these areas (Singh, Zinni, and MacLennan 2006).

However, most of these issues are clearly absent in the Portuguese science system, whose form of organisation contributes to the creation of a mobilisation pattern with different characteristics. First, business R&D, although showing significant growth in the past few years, does not yet provide a viable alternative in terms of employment for postdoctoral researchers. To a large extent, public funding remains the main promoter of scientific employment in Portugal. The fact that this funding is centralised in the FCT, which is the main distributor of scientific funding in Portugal, contributes to homogenise the working conditions among the different scientific areas. Research grant stipends and researchers' salaries are set nationally by the government and the FCT and working conditions are regulated by a group of professional statutes, such as the Research Grant Holders Statute or the Researching Career Statute, which include, among other things, a restriction of four weekly hours on the teaching workload for grant holders.

The importance of public funding for Portuguese research also contributes to level disciplinary science funding. The evolution of government funding more or less maintained a balance between the different disciplines. Although the budget values for the social sciences and humanities are among the lowest, the growth of science funding has maintained the same general distribution by scientific discipline.[6]

Thus, taking into account the available data for the national context, there seems to be no reason for a mobilisation following the same pattern of that of North America, primarily based on different features, in scientific terms and labour, of young researchers in different disciplines.

Since the associations are composed mainly of specific segments of professionals in the Portuguese scientific community, a possible explanation for the distribution by disciplines would be the very composition of these groups. When compared with the evolution of the number of PhD scholarships by subject area, the natural sciences seem to be over-represented among the members of the ABIC, while other areas such as engineering and technology, which constitute a substantial proportion of

---

[6] Source: FCT Conselho Directivo

the number of scholarships awarded each year, does not weigh significantly in the disciplinary composition of the association. The same thing can be inferred for ANICT, since the researchers from the natural sciences are clearly overrepresented—65% versus 41% of the total "Commitment to Science" researchers[7]—while other disciplines are underrepresented.

The literature provides other possible explanations for the concentration of the members of some disciplines. One would be that there are differences in the internal organisation and hierarchy of different disciplines which have consequences over the further deterioration of working conditions in certain scientific disciplines. Lefevre (2005) points out, for example, greater stratification, or an increase in inequality between junior and senior researchers, in the laboratory sciences, due to the greater dependence of research on access to costly scientific equipment. Nixon (1975) found different levels of acceptance of ideology as a factor for the mobilisation of young teachers in American unions. Among the technical disciplines (natural sciences and engineering) the ideological identification with the unions proved to be weaker and less frequent, leading to a reduced enrollment of young professionals from these disciplines in these organisations, while on the other hand, between the less technical disciplines (social sciences and humanities), with higher identification with these ideologies, the opposite tends to occur. None of these hypotheses seem to provide a sufficient explanation, since both are based on an opposition between the "hard" and "soft" sciences, which in the Portuguese case is insufficient to explain the differences observed in the data.

Another factor mentioned frequently as a promoter of mobilisation is the personal networks of the individuals themselves, that operate not only as channels for information flow, but also as influential relationships that encourage adherence and continuity of individual participation (Porta and Diani 2006; Kim and Bearman 2011). As can be seen (Table 8-5), professional contacts were one of the main channels for raising awareness of the association among prospective members, denoting the influence of professional relationships in promoting membership. This aspect becomes more interesting when we consider that co-worker mediation was the most important mean for promoting the associations among researchers in the natural sciences.

[7] Source: FCT, 2009, Mais cientistas para Portugal, Lisbon: FCT

Table 8-5: How members became aware of the association by scientific area (%)

	Natural Sciences	Health Sciences	Engineering and Technology	Social Sciences and Humanities
Through a co-worker	45.5	0	20.0	17.0
Through the association itself (website, event, newsletter)	24.8	66.6	40.0	38.3
Through a friend/family	1.0	16.7	13.3	14.9
Through the media	5.0	16.7	6.7	6.4
Through another source	23.8	0	20.0	23.4

N = 169

Thus, it is possible to affirm that the mobilisation for these associations had its genesis in groups with a strong presence of researchers from the natural sciences.

> When it [the association] started there was a great weight of the Faculty of Sciences, and the people of Biology and Chemistry, but now, (…) people from different areas begin to appear and we even have people from history. Fortunately, I think it is becoming more balanced in terms of scientific areas. (ABIC Interview)

> For example, in Minho that early core of researchers had at least two persons from the Social Sciences and Humanities [out of 23]. Here in Porto we have some people too... So, they are a minority. It [the association] is transversal, but it attracts members mainly from the areas of science and engineering. (ANICT Interview)

From this initial stage the way information about the associations spread throughout the community shaped their disciplinary composition. Having started with a predominance of natural sciences researchers, the associations' growth based on co-worker relations had an impact on maintaining the high proportion of researchers from these areas.

Expanding awareness of the association throughout the full universe of potential members is difficult and requires a substantial effort.

> …the greatest difficulty is to disseminate information about the association to the universe of potential members.. It's hard... because at this time the only way we have to get contacts is by hand. We go to Universities' websites and find who is a researcher, we try to get the e-mail and contact [and] request permission to use the e-mail for a mailing list (…) We feel we have the possibility of getting more members, and many researchers do not yet know the association. For example, we participated in an event and talked with some researchers who did not know the association (…) So this is a great difficulty at this time. (ANICT Interview)

The scientific literature on the motivations for joining associations is abundant and points both to instrumental factors and to more altruistic ones (Forsythe and Welch 1983; Smith 1994; Hwang, Grabb and Curtis 2005). Professional associations generally aim at long term objectives and can be heavily dependent on offering "selective incentives" to be able to capture the interest of potential members (Olson 1998). In this case, most members actually give less importance to utilitarian incentives such as obtaining information or benefiting from the associations' activities than to more idealistic ones (Table 8-6). As pointed out in other works on associative motivations (Dekker and Broek 1998; Gruen *et al.* 2000), "public goods", in this case the representation of interests and the contribution to the development of public understanding of science, can be as relevant to associative membership as incentives of utilitarian nature.

Regarding participation in these associations, the degree of commitment varies quite significantly. The most common form of participation is based on a very low level of commitment: most members just pay membership fees and read the information made available by the associations in their websites or publications. Other activities which require a little more involvement, such as voting and participating in events promoted by the association, are carried out by only half of the members. There is also a clear difference between these two sets of activities, which can be described as forms of a more passive participation, since they are mainly associated with the consumption of some "services" provided by the association, and the activities that require a higher commitment, involving a more active role. Activities such as volunteering for work within the association, being part of the governing bodies or writing for the associations' publications or website are carried out by a very small number of members (Table 8-7).

Table 8-6: Motivations to join by association*

	ANICT	ABIC
See my interest represented in politics	4.65	4.62
Belonging to a scientific community	4.29	3.89
Contribute to the development of science and society	4.13	4.22
Contribute to the public understanding of science	3.86	3.58
Meet people with similar interests	3.65	3.18
Obtain better knowledge about the scientific system	3.52	3.14
Obtain information about job opportunities	3.38	3.32
Obtain information about scientific events	3.30	3.05
Benefit from the association's activities	3.23	3.11
Obtain scientific information	3.22	3.09
N	69	182

* The values presented consists of the average value on a scale ranging from 1 (Strongly disagree) to 5 (Strongly Agree)

These observations coincide with descriptions of participation in other associations found in the literature, which report that non-professionalised associations, that is, associations that rely on the voluntary work of their members to maintain activity, are generally composed of: a group of nominal members, whose participation is almost limited to signing up for membership; a group of active members who participate in the activities; and a smaller group of committed volunteers who work in the organisation of these activities and in maintaining the organisational needs of the associations (Bekkers 2002; Freire 2004; Torpe 2003). This reliance on the commitment of a small group of members is also recognised by the associations.

> The governing bodies have fifteen people in total, but then there's more of an informal group, those who are part of the candidate list as substitutes. Altogether there are about twenty, twenty-two people, more or less, that are actively involved in the association. It is evident that in this group the level of involvement also varies. For example, mine is rather higher than that of the others who are not as involved. (ANICT Interview)

Table 8-7: Participation in the association's activities (%)

	ANICT	ABIC
Paying membership fees	94.3	97.3
Reading the association's publications or website/blog/forum	69.9	87.0
Participating in meeting/conferences	53.4	48.8
Recruiting new members	26.0	64.2
Voting in the elections	41.1	49.6
Participating in the general assembly	34.2	38.2
Writing for the association's publications or website/blog/forum	17.8	27.6
Doing voluntary work for the association	11.0	21.1
Being part of the governing bodies	12.3	13.0
Coordinating a section/publication	4.1	13.0
Making donations to the association	2.7	4.9
N	73	123

Conclusion

The characterisation of ABIC and ANICT members reinforces the idea that these associations are permeated by professional subdivisions within the scientific community, with each association linked to people in distinct stages of the scientific career. ABIC is, as expected, mainly composed of grant holders. ANICT, in turn, is composed primarily of research fellows, mainly contract researchers from the "Commitment to Science Programme". The constitution of the associations also reflects some characteristics of the scientific and technological system, in particular the concentration of the recent growth of human resources in R&D in universities and in the two largest metropolitan areas of the country, Lisbon and Oporto.

The more interesting aspect seems to be the disciplinary distribution of association members, with an overrepresentation of researchers from the natural sciences. This feature runs contrary to the trend found in the North American context, where the young researchers' movement is strongly rooted in the social sciences and humanities. This difference, or rather the specific disciplinary concentration found for the Portuguese case, without disregarding other possible factors, seems to derive from: 1) the lack of differentiation in working conditions between the social sciences and humanities and the other scientific disciplines that more easily attract external funding, as a result of a heavy reliance of the science system on public funding; and 2) the fact that the associations were founded by groups with a majority of members from the natural sciences, which combined with the recruitment of new members through professional contacts, maintained the predominance of the members from these scientific areas.

Most members of the associations tend to join them for the "public goods", such as benefits of the group or for science and society and general. Individual benefits seem to have a lesser impact as a motivation for membership.

In terms of participation, these associations are highly dependent on the voluntarism of a few more active members, who are responsible for organising most of the activities and producing most of the information that will be "consumed" by the less active members.

Bibliography

Bekkers, R. 2002. "Participation in Voluntary Associations : Resources, Personality, or Both ?" In *5th Conference of the European Sociological Association,* Helsinki.

Clawson, D., & M. A. Clawson. 1999. "What has happened to the US labor movement? Union Decline and Renewal." *Annual Review of Sociology* 25: 95–119.

Dekker, P., & A. Van Den Broek. 1998. "Civil Society in Comparative Perspective : Involvement in Voluntary Associations in North America and Western Europe" 9 (1): 11-38.

Ehrenberg, R. G, D. B Klaff, A. T Kezsbom, & M. P Nagowski. 2004. "Collective Bargaining in American Higher Education." In *Governing Academia,* ed. R. Ehrenberg, Ithaca: Cornell University Press: 209-233.

Forsythe, DP, & S, Welch. 1983. "Joining and Supporting Public Interest Groups: A Note on Some Empirical Findings." *The Western Political Quarterly* 36 (3): 386–399.

Frege, C. M., & J. Kelly. 2003. "Union Revitalization Strategies in Comparative Perspective." *European Journal of Industrial Relations* 9 (1) (March 1): 7-24.

Freire, J. 2004. *Associações Profissionais em Portugal.* Edited by João Freire. Oeiras: Celta.

Gruen, T. W, J. O. Summers & F. Acito. 2000. "Relationship Marketing Activities, Commitment, and Membership Behaviors in Professional Associations" *Journal of Marketing* 64 (July): 34-49.

Hoeller, K. 2007. "The Future of the Contingent Faculty Movement", *Inside Higher Education*, (November): 1-6.

Hwang, M., E. Grabb, & J Curtis. 2005. "Why Get Involved? Reasons for Voluntary-association Activity Among Americans and Canadians." *Nonprofit and Voluntary Sector Quarterly* 34 (3): 387-403.

Julius, D. J., & P. J. Gumport. 2002. "Graduate Student Unionization: Catalysts and Consequences." *The Review of Higher Education* 26 (2): 187–216.

Kaplowitz, M. D. 2004. "A Comparison of Web and Mail Survey Response Rates." *Public Opinion Quarterly* 68 (1) (March 1): 94–101.

Kerchner, C.. 2003. "Knowledge Workers : Trade Unionism's New Frontier." *Higher Education* (11): 11–18.

Kim, H., & P. S. Bearman. 2011. "The Structure and Dynamics of Movement Participation." *Structure* 62 (1): 70–93.

Lafer, G. 2003. "Graduate Student Unions: Organizing in a Changed Academic Economy." *Labor Studies Journal* 28 (2): 25–43.

Lefevre, W. 2005. "Science as Labor." *Perspectives on Science* 13 (2) (June): 194–225.

Meyer, D. S. & S. Staggenborg. 2011. "Movements, Countermovements, and the Structure of Political Opportunity." *Structure* 101 (6): 1628–1660.

Nixon, Howard L. 1975. "Faculty Support of Traditional Labor Tactics on Campus." *Sociology of Education* 48 (3): 276.

Olson, Mancur. 1998. *A Lógica Da Acção Colectiva*. Oeiras: Celta.

Porta, D. & M. Diani. 2006. *Social Movements. An Introduction*. Malden, Oxford & Carlton: Blackwell.

Rhoades, Gary, and Robert a. Rhoads. 2002. "The Public Discourse of U.S. Graduate Employee Unions: Social Movement Identities, Ideologies, and Strategies." *The Review of Higher Education* 26 (2): 163–186.

Rhoads, R. A. & G. Rhoades. 2005. "Graduate Employee Unionization as Symbol of and Challenge to the Corporatization of U.S. Research Universities." *The Journal of Higher Education* 76 (3): 243–275.

Sax, L. J, S. K Gilmartin & A. N, Bryant. 2003. "Assessing Response Rates And Nonresponse Bias In Web And Paper Surveys." *Research in Higher Education* 44 (4): 409–432.

Sheehan, K. & M. Grubs. 1999. "Using E-mail To Survey Internet Users In The United States: Methodology And Assessment." *Journal of Computer-Mediated Communication* 4 (3).

Sheehan, K. B. 2006. "E-mail Survey Response Rates : A Review." *Communication*, 6(2): 1–30.

Singh, P., D. M. Zinni & A. F. MacLennan. 2006. "Graduate Student Unions in the United States." *Journal of Labor Research* 27 (1) (March): 55–73.

Smith, D.H. 1994. "Determinants of Voluntary Association Participation and Volunteering: A Literature Review." *Nonprofit and Voluntary Sector Quarterly* 23 (3): 243–263.

Torpe, L. 2003. "Democracy and Associations in Denmark: Changing Relationships Between Individuals and Associations?" *Nonprofit and Voluntary Sector Quarterly* 32 (3) (September): 329–343.

Zhang, Y. 2000. "Using the Internet for Survey Research: A Case Study." *Journal of the American Society for Information Science* 51 (1): 57–68.

CHAPTER NINE

NEW (?) ACTORS IN THE PUBLIC COMMUNICATION OF SCIENCE: THE ROLE OF SCIENTIFIC ASSOCIATIONS

CRISTINA PALMA CONCEIÇÃO AND ANA DELICADO

In the twentieth century, the critical role scientific knowledge plays in the processes of economic, social and cultural change became clear. The presence and significance of science in the most diverse spheres of contemporary life was already an indisputable fact (Stehr 1994; Beck, Giddens and Lash 2000; Castells 2002). However, the need to bring science closer to other social institutions remains, or is even increasing (Miller *et al.* 2002), as evidenced by many public policies aimed at consolidating the so-called 'knowledge society', whether focused in the economic (e.g. strengthening links between universities and business), educational (e.g. the enhancement of science and technology education, namely through experimental learning), political (e.g. the engagement of experts in public decision-making) or cultural (e.g. the creation of new science museums or the improvement of science in the media) realms. However, even if science is assumed to be a resource that can be of service to society, science is also increasingly seen as a social problem, since the processes of incorporating it into society are often not free from inequalities, uncertainties and controversies (Habermas 1970; Irwin and Wynne 1996; Costa, Conceição and Ávila 2009; Dierkes and Grote 2000).

Therefore, it is not surprising that so much attention is increasingly being paid, in many national contexts, to the issue of public communication of science, including a wide range of initiatives being carried out by an array of science-related professionals (Bauer *et al.* 2007).

This chapter seeks to explore the role scientific associations have come to play in this movement. Even if these organisations are far from being the only ones engaged in it, several authors have identified them as privileged actors in building bridges between science and society (Rogers 1981; Hin and Subramanian 1999; Miller *et al.* 2002; Evans 2010) and some even point to their comparative advantages: the ability to mobilise scientists and engage them in civic action; an added legitimacy stemming from their collective nature and their greater neutrality in view of particular interests of universities and research institutions. Yet, to what extent is communication with non-specialised publics among their aims? And, if so, what type of activities are they carrying out?

Scientific associations in historical perspective

There are not many studies focused on contemporary scientific associations–here defined as private non-profit organisations in science not (solely) dedicated to research–thus, little is known about the current activities of such organisations. Existing literature mostly adopts a historical perspective, describing and discussing the role that early scientific societies, such as the Royal Society (England, 1660) and the Académie des Sciences (France, 1666), had on the rise of modern science and in the dissemination of this new form of knowledge (Merton 1938; Shapin 1996).

Among the activities of these societies were public lectures and scientific demonstrations that gathered together people from very different backgrounds (Rasse 2002; Raichvarg 2005; Zilsel 2000). From this perspective, it may be considered that scientific societies and their members were among the earliest public communicators of science. It is, however, important not to forget that in those days the distinction between scientists, other intellectuals and amateurs was still rather tenuous (Ben-David 1984; Dear 1985) and that those public meetings were not just a stage for presenting and debating the new discoveries and methods of experimental science, but were also a space for validating such knowledge.

This panorama has changed with the professionalisation and specialisation of scientific production and science education. The distinction between experts and lay people has increased significantly (Bensaude-Vincent 2001) and an array of new functionally differentiated institutions emerged, some more strictly focused on research, others on education, dissemination, etc. The early science academies did not disappear, but their mission has certainly changed.

By the nineteenth century, a new type of scientific association emerged: the disciplinary societies. These were closely related to the rise of new scientific fields and the search for new platforms for sharing and certifying knowledge among peers (such as through specialised scientific journals). The number of such scientific associations grew substantially, as did the degree of specialisation and internationalisation. Most became more focused on providing services to their own members (experts in the field), rather than to the public. Even if some kept a number of science communication activities for non-specialists, it was considered to be a less important task (Rilling 1986; Chaline 2002; Schofer 2003).

Meanwhile, still in the nineteenth century, some countries witnessed the rise of another type of association: national transdisciplinary scientific associations, dedicated to the "advancement of science", such as the British Society for the Advancement of Science (created in 1831) or the American Association for the Advancement of Science (created in 1848). They were intended to be a forum for communication among scientists, but also a way to draw national attention to science and promote its development. Throughout the twentieth century, many of these played an active role in the public dissemination of science, often seeking the attention of policy makers, the media and other publics (Rogers 1981; Teich 2002; Gascoigne *et al.* 2010; Bodmer 2010).

In the last decades of the century, some of these organisations played a decisive role in launching the new movement for the public understanding of science, showing concern about the relationship between science and the public and a willingness to improve their work in this area (Gregory and Miller 1998; Miller 2001). The famous report by the Royal Society (1985), diagnosing the risks of increasing levels of ignorance about science and mistrust of its institutions and professionals, and thus stressing the urgent need for engaging a wide range of players in the efforts of raising public awareness and public debate on scientific issues, was of unquestionable importance in this matter.

Nevertheless, some questions are still to be answered. Are all types of scientific associations really participating in this movement? And, if so, are they all using the same methods?

Methodology

This chapter results from a broader research project on Portuguese scientific associations, conducted between March 2010 and August 2012.[1]

---

[1] SOCSCI Scientific Societies in Contemporary Science, funded by the Portuguese Foundation for Science and Technology (PTDC/CS-ECS/101592/2008), conducted at the Institute of Social Sciences of the University of Lisbon, with the collaboration of the Centre for Research and Studies in Sociology (University Institute of Lisbon) and SOCIUS (ISEG - University of Lisbon).

Although the Portuguese situation may have some particularities, focusing on this case can also be fruitful for several reasons: the small size of this scientific system makes a thorough study of its associations possible; its high degree of internationalisation may allow some generalisations to other national contexts; the strong investment the government has made in recent years in science and science dissemination provides a strategic moment to understand the challenges scientific associations are facing at a time of transition towards a knowledge-based society.

The first stage of this study comprised a census of scientific associations in the country, since there was no organised or reliable source to identify them. A broad definition of scientific associations was devised, comprising all types of non-governmental, non-profit organisations that met one or more of the following criteria: calls itself a scientific association or society; has scientific aims in its mission statement; carries out science-related activities (e.g. research, funding, communicating, disseminating, regulating science or representing scientists' interests). Using a variety of sources, a total of 363 scientific associations were identified and their primary data compiled. A set of quantitative and qualitative methodologies were then applied in order to provide an in-depth understanding of the current roles of these organisations (concerning several issues).

The empirical evidence for this chapter stems mostly from: an online questionnaire survey of the Portuguese scientific associations (response rate 32%); an in-depth study of 24 associations, including interviews with presidents (and, in some cases, other board members), document analysis (of reports, statutes, regulations, websites, publications, etc.), a questionnaire survey of members, and ethnographic observation at events; a workshop for debating the preliminary results with association representatives; an online questionnaire survey of a sample of researchers working in Portugal (about their associative affiliations and practices).

Based on the data collected through this research, a typology of scientific associations has been created, according to their different purposes. This typology comprises three "ideal types" (whose boundaries are, nonetheless, fairly flexible):

- Disciplinary scientific societies: representing 73% of the Portuguese scientific associations, they are mainly focused on the promotion of a scientific discipline and tend to assume communication among peers as their major purpose (namely the diffusion of research results through specialised journals and conferences). The vast majority of their members are academics, even if some of them also gather other practitioners or students. Examples of this type of association are the Chemistry Society or the Sociological Association. The oldest scientific associations are also found in this group.

- Science dissemination associations: this group is comprised of organisations such as astronomy clubs, nature conservation associations, archaeology groups and associations for the diffusion of robotics or new information and communication technologies, etc. Although the majority are disciplinary, there are some cases of transdisciplinarity. They represent 22% of the total and are likely to be more socially oriented than the other two types. They focus mainly on the public understanding of science and their members are usually both professional researchers and amateurs of science (see Table 9-2).
- Professional associations of scientists: being the smallest group (representing just 5% of the total), this type consists of associations mainly focused on the representation of the interests of science and engineering professionals (researchers predominating). They can be either disciplinary (e.g. associations of geologists or biochemists) or transdisciplinary organisations (e.g. trade unions of university teachers or associations of grant holders) (see the previous two chapters in this book).

A growing number of science dissemination associations

Overall, the growth of scientific associations in Portugal is a fairly recent phenomenon. Although the oldest scientific societies date from the nineteenth century, close to 90% of associations identified in the census were founded after 1970, with slightly more than one-third established in the last decade alone. In the absence of up-to-date international data, it is difficult to assess how common this growth rate is. However, considering the Portuguese context, it is easy to identify the significant factors that help explaining it. In brief, complete freedom of association was only granted in 1974, after the democratic revolution; in terms of science, the national scientific system has only experienced significant growth since the 1980s. This development has brought opportunities, such as accrued "critical mass" and specialisation, but also threats, namely an increased competition for resources, and needs, such as the need to strengthen the communication between peers across the country, the dialogue of scientists with policy makers and other stakeholders, and public support for science. All these trends are likely to have favoured the emergence of scientific associations.

For the purpose of this chapter it is particularly interesting to note the growth of the science dissemination associations. These were virtually non-existent until the 1980s, but have since increased non-stop and are now no longer a negligible proportion of the scientific associations operating in the country (Table 9-1). The emergence and growth of this specific type of organisation has been strongly influenced by the rise of "scientific culture", both as a scientists' concern and as a policy priority in Portugal (Gonçalves and Castro 2002).

Table 9-1: Scientific associations in Portugal by year of foundation

	Disciplinary scientific societies	Science dissemination associations	Professional associations of scientists	Total
Until 1950	23	-	-	23
1951 - 1960	6	-	-	6
1961 - 1970	8	-	-	8
1971 - 1980	23	5	2	30
1981 - 1990	52	6	3	61
1991 - 2000	45	27	3	75
2001 - 2011	67	26	4	97

Source: Census of scientific associations in Portugal, 2011. Data unavailable for 60 associations.

Some of these associations were actually formalised in order to participate in initiatives promoted by the Ciência Viva Agency–the National Agency for Scientific and Technological Culture. Created in 1996 under the auspices of the Ministry of Science and Technology, the Ciência Viva programme (which resulted in the Agency) has been the main catalyst for the public communication of science in Portugal and, consequently, for the emergence of many associations specially committed to this issue. Although associations tend to be generally understood as crystallised forms of collective action that emerge from civil society, this case shows an interesting example of a new social movement that has been strongly supported by governmental stimulus (Costa, Conceição and Ávila 2009).

> By that time the Ciência Viva agency invited us to carry out some "Astronomy in the summer" activities, namely astronomical observation sessions in the countryside, in areas where such initiatives tend to be rare.

> That's how we started doing science communication within the country and created the Astronomer's Portal. We were about twenty astronomers, working at the Observatory as masters, PhD students, and postdoctoral researchers. Some amateur astronomers also joined us, and we set up the association (...), a non-profit association, aimed at science dissemination, mainly astronomy. The Agency demanded that we had some sort of legal status. (Interview with the president of the NUCLIO Interactive Astronomy Nucleus)

As regards disciplinary fields, most of the science dissemination associations are dedicated to the natural sciences, the exact sciences and, in some cases, the humanities. The appearance of associations in the first two scientific fields is probably due to the Agency favouring activities in these disciplines, whereas the associations in history or archaeology tend to be part of a longer tradition of civil society initiatives concerning cultural heritage. There are only a few associations of this type focused on engineering or the social sciences.

Some of the associations primarily devoted to the public communication of natural science issues also have the status of "environmental non-governmental organisation" (a status that allows them to access other funding sources) and they usually include nature conservation among their main objectives.[2]

> The association main aims are: to promote the study and conservation of birds in general and raptors in particular, as well as their habitats; to disseminate and promote scientific culture and science education, having: a) knowledge and scientific methodology as technical support, in order to promote 'citizen-scientist', b) the birds and the natural heritage as the locus of activity (Statutes of the Scientific Association for the Conservation of Birds of Prey).

> A group of friends founded the association in order to do some research studies and to promote ornithology. (...) It began as a purely amateur association with the contribution of researchers, ornithologists and bird watchers as well. Meanwhile, working groups were also formed, as were regional and thematic groups, and the association became bigger and bigger. (...) I would say that the mission remains the same: promoting ornithology, and the conservation of birds and habitats. Fortunately we are now reaching much more people than just the scientific community. (Interview with the president of the Portuguese Society for the Study of Birds).

---

[2] On the affinity between the environmental movement and the public communication of science, see Yearley 2008.

A few other associations are classified as "non-governmental organisations for development", such as Scientists in the World, created in 2007 with the aim of promoting science dissemination in "third world" countries and of teaching local communities how to use simple technologies to meet some basic needs.

> The idea was to work at the intersection between science and development cooperation. (...) We are all scientists with no experience in the world of cooperation; but several of us, in more or less private visits, or in volunteering in developing countries (Mozambique, Timor, etc..), had come across the problems that exist in these countries... and it's hard not to be touched. In my particular case, I visited Timor and went back thinking about what a theoretical and useless scientist can do to help that country (...). So we thought about education projects and about the introduction of small technologies. (Interview with the president of Scientists in the World).

As previously mentioned, there are also science dissemination associations not devoted to any specific scientific field. Such is the case of VAC "Live Science" Association, an association created in 2005 by a group of young researchers who wanted to raise public awareness of science and carry out fundraising activities in Portugal, such as they had seen whilst studying and working abroad.

> It all began when a number of scientists returned from abroad and realised that there were many things still missing in the country as regards science dissemination. In other countries, they were giving more money to science, people were more knowledgeable about science, science was being included more in the newspapers, on television, and all this in a way that permeated society far more than here. So they tried to create an association that is, as they say, grassroots (…) to explore something that we thought that was absent, and still is: private investment in science (…). We would like to have a platform allowing us to do two things. One is to increase public awareness, leading Portuguese society to recognise, to simply know, who their scientists are, what are they doing, where are they, and how science can contribute to the advancement of society, to education... Get citizens closer to science, to scientists. We thought that, by doing so, we would raise awareness among citizens, the private companies, etc. in order to get them to participate in their country's research. Since not all of us can do science experiments, they might give one euro, two euros, some money, which enables the creation of an alternative source of funding to the State. That was the big idea. And we thought that the easiest way of doing so was to create an association. (Interview with the executive director of VAC)

Therefore, even if the birth of many of these scientific associations clearly has something to do with recent public policies, the excerpts presented above also reveal a more grassroots social movement that seems to primarily result from the researchers' wish for a closer connection between science and society.

Although most of these organisations are essentially formed by scientists (many of whom are in search of social support for their activities), some also include "non-scientists" among their members. These amateurs often play an important role in the life of the associations. They are not simply the audience for science communication activities; some actively collaborate in these activities as facilitators in the approach to local communities, and some are even engaged in research projects carried out within the framework of associations (see the next two chapters in this book). Thus, these cases can be understood as an illustrative example of the emergence of new forms of knowledge production/dissemination (i.e. citizen science).

> The professional scientists collaborate with us, for example, in the review of scientific contents. The amateur astronomers who are members of the association are also a precious help. They support many of the observation sessions [for various audiences] and they help teachers to adapt the resources to schools. Teachers also collaborate... and there's also IT experts collaborating with us. (Interview with the president of the NUCLIO Interactive Astronomy Nucleus)

Table 9-2: Proportion of "non-scientists" among the Portuguese scientific associations' members (%)

	Disciplinary scientific societies	Science dissemination associations	Professional associations of scientists	Total
50% to 75%	-	7.7	-	2.3
25% to 50%	2.1	38.5	-	12.8
Up to 25%	25.0	34.6	8.3	25.6
None	72.9	19.2	91.7	59.3

Source: survey of the Portuguese scientific associations, 2011; N = 86; *Kruskal-Wallis test: p = 0.000

Is public communication of science an issue for all types of scientific associations?

One might think that the relatively high number of new science dissemination associations in the country would be a sign that other types of scientific associations have not been so interested in science communication activities. That does not seem to be the case, at least not in many disciplinary scientific societies.

As expected, science dissemination associations tend to be engaged in science communication activities for non-specialised publics more often. According to the survey conducted in Portugal in 2012, 72% of this type of association performs this type of activities regularly, including both events for young people and for the general public (Table 9-3). Nonetheless, many disciplinary scientific societies also claimed to be involved often in such actions. This participation tends to be more common in the case of activities aimed at the general public: about 50% reported doing it regularly; less than 10% reported never doing so. Younger audiences tend to be addressed by a smaller number of these associations, and less frequently; however, they are far from being neglected by many of these organisations.

Table 9-3: Frequency of public communication of science activities, for young people and the general public, by type of association (%)

	Disciplinary scientific societies	Science dissemination associations	Professional associations of scientists	Total
For young people*				
Regularly	16.4	71.9	22.2	34.3
Occasionally	38.2	18.8	38.9	32.4
Never	45.5	9.4	38.9	33.3
For the general public**				
Regularly	50.9	71.9	16.7	51.4
Occasionally	40.0	15.6	55.6	35.2
Never	9.1	12.5	27.8	13.3

Source: survey of the Portuguese scientific associations, 2012; N = 105;
*p = 0.000; Cramer's V = 0.377; ** p = 0.003; Cramer's V = 0.279

For some of these disciplinary scientific societies, engagement in the public communication of science is not something new. Nevertheless, the growing interest in these activities shows a certain change in the understanding of the mission of many of these entities. Given the internationalisation of science and the increasing importance of transnational forums for peer communication (which are threatning the relevance of national conferences and scientific journals) many have come to redirect their actions towards other (non-scientific) publics. By communicating with broader audiences, the associations aim to contribute to the promotion of the scientific culture of the Portuguese population–intending primarily to reinforce their interest in such topics and their appropriation of science (as regards intellectual valorisation, solving practical problems, or even the empowerment of civic participation in science-related debates). However, obviously, there are also other (more corporatist) aims in such actions–to strengthen the public legitimacy of science, the support for public funding of science, and the interest of young people in pursuing science in higher education.

> Many people discovered—especially in the area of biology, I think—that public communication of science turns out to be a way of valuing their field of knowledge. (...) Some people do it because they like it and they believe in it, but there are plenty of people who are doing it because it ends up being a weapon, a form of activism. Valuing the knowledge they produce will give them visibility; it will be used for safeguarding the area. (Statement by the representative of the Grant Holders Association at the workshop)

The survey of scientists (concerning membership of scientific associations) does indeed show that, although public communication is not among the missions that scientists identify as most critical, many of them recognise it as an important and relevant area of intervention for these entities (Table 9-4). Many also mentioned that their personal interest in participating in science outreach activities constitutes one of the main reasons why they decided to become a member of the association (62% in the case of members of science dissemination associations, 26% in the disciplinary societies). These figures give a favourable view of the possibility of these organisations becoming a privileged space for scientists to interact with other social actors.

Despite also recognising the relevance of these initiatives, the professional associations of scientists tend to concentrate their efforts in other areas. There are only a few cases of awareness-raising events for young people (mostly about scientific occupations) carried out by this type of association.

Table 9-4: Researchers' motivations for belonging to a Portuguese Scientific Association, by type of association (%)

	Disciplinary scientific societies	Science dissemination associations	Professional associations of scientists
Contribute to the promotion of scientific culture in society	44.7	86.8	33.5
Enjoy participating in public communication of science	25.5	61.8	14.7

Source: Survey of researchers working in Portugal, 2012; N = 462.

Table 9-5: Collaboration with other institutions for the purpose of public communication of science, by type of scientific association (%)

	Disciplinary scientific societies	Science dissemination associations	Professional associations of scientists	Total
Ciência Viva Agency*	15.8	70.4	16.7	35.1
Science museums**	18.4	44.4	16.7	27.3
Basic/Secondary schools***	28.6	79.3	53.8	51.9
Municipalities****	34.3	72.4	46.2	50.6

Source: survey of the Portuguese scientific associations, 2012; N = 77;
*p = 0.000; Cramer's V = 0.544; ** p = 0.045; Cramer's V = 0.284; *** p = 0.000; Cramer's V = 0.461; ****p = 0.009; Cramer's V = 0.348

The engagement of scientific associations in the public communication of science is likely to be favoured by the collaboration with other institutions, mostly with the previously mentioned Ciência Viva Agency, but also with science museums, schools and municipalities (Table 9-5). Sometimes these institutions seek the help of scientific associations to develop their own science outreach initiatives; in other cases they merely provide some support (financial, logistical or advertising) for the associations' public events.

> We have some support... some of our initiatives are the result of partnerships. This initiative is a specific action that began inside the SPEA,

the Ciência Viva joined later. It is called "An eye on the birds" and it is a project aimed at anyone who wants to watch birds during weekend walks. (...) Maybe Ciência Viva realised that this was important... so we have had this partnership for some years, and we want to keep it. (...) The willingness to cooperate is considerable, from both parties... but then, sometimes, there are also circumstances that may constrain things... (Interview with the president of the Society for the Study of Birds)

Usually those in demand are schools or municipalities... There are many entities who contact us because they want to do something (the Culture House, for example...). The municipality of Cascais is one with whom we collaborate more regularly. There's always something on, as there is a formal agreement with the municipal authorities. But we support many other municipalities. And sometimes we are the ones searching... for example, we are now about to do an action in Pampilhosa da Serra, which was an initiative of ours. (Interview with the president of the NUCLIO Interactive Astronomy Nucleus)

As regards media contacts, associations tend to have a more passive role. Most association representatives state that they try to reply to requests from journalists in search of researchers willing to comment on or support the production of science-related news. The media are seen by the associations as an important vehicle to reach broader audiences–often accessed by means of press releases and occasionally sought for other forms of collaboration, such as the co-production of television or radio programmes–even if most of the interviewees ended up complaining about the treatment of scientific issues in the media.

What kind of activities are they doing?

In what concerns scientific dissemination for young people, associations are often contacted by schools asking for experts to deliver lectures (or, more rarely, scientific demonstrations); thus, they frequently end up working as intermediaries between young people, teachers and members of the scientific community. Some associations already have a list of people with some experience and interest in this type of event. Nonetheless, some also stress the difficulties in raising this type of collaboration among their members.

We have information sessions aimed at secondary education students. These students hesitate on which field of study they will choose at university. We have had requests from schools and we also offer our services. We have been asked to do information sessions on what is

> biochemistry, where can you get a degree in biochemistry, (...) what does a biochemist do, what do we study, what are our areas of expertise. (Interview with the President of the Biochemists Association)

> We have no chance of running about, all over the country. We help whenever we can, but it is not a systematic thing. (...) People who are working in a hospital don't have much time to do it. But we try to ask people who have more to do with the things we have been asked to talk about, and who might have more time and willingness to do it. Then it's up to them. (Interview with the President of the Portuguese Neurology Society)

Nevertheless, some associations take the initiative and involve students in their own dissemination activities. For instance, the Neurological Society is in charge of the activities of the International Brain Awareness Week, during which its members deliver lectures at schools, participate in roundtables, exhibitions and seminars for the general public, and students are also invited to visit research laboratories.

In close collaboration with science educators, some scientific societies are also well-known for being responsible for organising competitions for students, generally designated "Olympiads". This occurs in "traditional" science subjects (mathematics, physics, chemistry), but is expanding to a growing number of fields of knowledge (biology, computer science, philosophy) and, in some cases, is acquiring an international scope.

Scientific societies, such as the Biochemistry Society, even make room in their annual conferences for sessions aimed at high school students, who are allowed to attend other sessions and to visit the posters section, in what is clearly an effort to attract potential undergraduates to university degrees in this field.

The idea of opening the doors of the professional science sites for students to have a "first look" also lies behind the organisation of visits to laboratories and other places where scientists work. To arrange these events some societies seek the support of the research units where their members work, but some science dissemination associations are able to present and involve the public in their own research projects, many of which follow the practices of "citizen science" (Bonney *et al.* 2009; Clark and Deborah 2001; Conrad and Hilchey 2011; Greenwood 2007; Hartman 1997; Leach *et al.* 2005) or "open science" (Hetland 2011; Jankowshi 2007), as mentioned above.

> This is a good way for us to teach students (as well as some of our members, and other publics) some of the issues we deal with... so that people have a different understanding of the planet, and so on... I think

> that, this way, these things (climate change, the impact climate change has on the planet) can be better taught and debated, by seeing things in practice. (...) These things are difficult to explain when we make a powerpoint... it's different being actually in the field. (...) During the science and technology week, we have always opened the doors of the laboratory to the general public. And even other occasions, if there is a group of 10 or 15 people asking us to show the collection, we are glad to do it. Soon we are going to receive a group of 30 people from a nearby village who have asked to see these bones. They will see us working, experience how we clean and classify the bones (...). We really think that the public's closer relationship with the palaeontology community will bring enormous savings in heritage preservation (...). Additionally, these events also reinforce the researchers' feeling that they are doing interesting and useful things. (Interview with the President of the Leonel Trindade Natural Science Society)

Several of such science dissemination associations are making a point of engaging students, not just by talking and showing things, but also by encouraging the young public to experiment with scientific procedures directly (e.g. through inquiry-based science learning projects, often carried out with the help of teachers and sometimes also involving the parents). Some of these projects are part of national or international initiatives, others have strong local roots. For instance, the Society for the Study of Birds has a nationwide activity, which includes the participation of research institutes, universities and museums, called "In a wood near you", in which school children collect information on forest ecosystems (fauna, flora and human threats) and upload it to a website. This association also has other initiatives that try to combine science teaching (using real science experiments) and community engagement in biodiversity conservation in sites where the organisation is developing research/conservation projects. Another Portuguese association is coordinating a world-wide project (the Galileo Teacher Training Program) aimed at training teachers to help students to make astronomical observations, via the internet, through telescopes located all over the world. Some of the students have already gathered information that can be included in databases that are accessed by both professional astronomers and amateurs. Another interesting example is the work done by the association Scientists in the World with schools in developing countries: for example, an atelier about the physics phenomena underlying local musical instruments built by the students, or a project on how to build a solar kitchen oven that could be of use for communities living in remote villages.

> We are not well-known by the general public… but, in the areas where we are operating, we are certainly well-known. For example, in S. Miguel, our centre does environmental education programmes with the teachers from all schools on the island. We are working with them in order to reach the students and then the local community as a whole. (…) We also had an initiative that involved several schools across the country, namely rural areas (…) Children love these things, they love doing practical work and getting engaged in extra-curricular activities. And it's easier to talk with children, and also great fun for us. (Interview with the president of the Portuguese Society for the Study of Birds)

> Astronomy is one of those scientific fields in which we don't have to build a school laboratory in order to get the students to do practical work. (…) An example: we asked the teachers to involve their students, next April, in astronomy dissemination activities aimed at local communities. The children are going to be in charge of preparing the telescopes so that their parents and grandparents can come to the school to watch the sky with them. (Interview with the president of the NUCLIO Interactive Astronomy Nucleus)

Many of these projects may be seen not just as a way of promoting science education among the youngsters (using more thanmerely discursive approaches), but also as a way of reaching adult audiences through children. On the other hand, the same sort of hands-on and contextualised approach is sometimes being used to capture the interest of exclusively adult publics, as in the case of having experts from the associations helping non-specialists to collaborate in the censuses of natural specimens, or monitoring public visits to parks, archaeological sites, agricultural fields or sites for bird watching.

> We have also other type of events, aimed at the general public... the wine sector visits, for instance (the next will be the 12th edition). (...) We visit wineries, horticulture companies... Sometimes, part of the day is devoted to a more strictly techno-scientific conference, and then we have an open visit. (...) In the past, visits to gardens and parks were also organised. (...) It's very funny because it attracts a different audience. Some had nothing to do with horticulture, but became members of the association just to go on those visits. It was an idea inherited from the previous board. At the beginning I didn't pay much attention to it, but then I've realised that people were really interested. (Interview with the president of the Portuguese Horticulture Association)

Nevertheless, in general, public talks are still the most common method used by Portuguese scientific associations (disciplinary scientific societies in particular) to address the public (whatever their age). In an attempt to capture the interest of broader audiences (not just the ones who are already very interested in science topics and willing to listen to scientists talk in their own specialised dialects), some promoters opt to focus the talks on particularly fascinating stories of science or on current applications of science. Sometimes they also try to give people the opportunity to debate these issues, so that the meetings do not end up being just like a class. Occasionally, they choose to hold the sessions outside schools, universities or other formal sites, in an effort to give these events a more informal ambience and to attract new publics (such as science cafés).

Other forms of public communication of science adopted by some associations (mostly by the ones more strictly devoted to science dissemination) are exhibitions, popular science books and newsletters, and, more rarely, theatre plays and stand-up comedy shows. Many association representatives also mentioned the role the internet may play as an instrument for reaching broader audiences. Social network profiles, online videos, portals and blogs are all communication methods that are more frequently used by some of these organisations. However, the need to keep web content updated is often mentioned as a problem because it requires human and technical resources not always available in these associations, as the majority heavily depends on the work of volunteers.

> For example, we did an exhibition at LX Factory… Nobody goes there to see science stuff. People go there to have fun, to see art exhibitions… But, the same way people can see an exhibition about any other topic, they can see photos that have been made by scientists and that show scientific images, and they end up learning something with it. Or, as regards theatre, we can do stand-up comedy with scientists and about science. That's our idea: to reach publics that we normally don't reach. Of course, the things we do always have something to do with our own interests. I'm personally interested in images, so I took photos and made documentary films. David likes theatre and humour, so he joins his love of science with it… People come with their own ideas, and then we see if we can do it or not… On our website we have "we are looking for new ideas!" (…) Raquel is a great storyteller, so she thought of this project, the "Sea's little girl". (…) This was a fantastic and very enriching experience, with children with leukaemia, at the Portuguese Oncology Institute. For three months, once a week, she went there with this story, to get them to know more about sea life. She brought some fish and other marine specimens and explained to the children some of their characteristics and then we asked them to express by painting the things they had learned, the things they liked the most, the facts they thought it would be funnier to explain. These paintings were then sold at an auction; the children were also there, so they felt quite

> proud! And the money we collected was used to award a scientific prize for cancer research (Interview with the executive director of VAC).

> We created an online platform to answer people's questions about ecology. (...) Yesterday I received a question from someone who wants to plant trees at a certain location and wants to know which are the most suitable plants to put in that ecosystem in order to produce honey. So we redirected this request to the right person (we have a wide network in the field of ecology), the expert usually answers and then we give feedback to the one who asked. (Statement by the representative of Portuguese Ecology Association at the workshop)

Using the potential of the internet, associations can also function as a channel to convey the needs and expectations of citizens and civic groups to the scientific systems. This role is already being played in other countries by means of science shops (Leydesdorff and Ward 2005). Some Portuguese scientific association leaders are aware of this trend, and some are already implementing online platforms for people to ask questions to scientists. However, currently, none has been able to put such an experience into operation on a more regular and formal basis.

Final remarks

The vast majority of the Portuguese scientific associations seem to recognise the importance of bringing science closer to the public. Although for many (namely professional associations of scientists, but also many disciplinary scientific societies) this task is far from being regarded as their main mission, many are already showing a willingness to participate in this movement and some are even becoming fairly active players in this field.

The emergence, in recent years, of a considerable number of associations primarily devoted to science dissemination is certainly a good indicator of these trends, whether this is a direct result of public policies concerning the issue of the "scientific culture" of the Portuguese population or a sign of the scientists' growing awareness of the need to establish a closer relation with society (in search of public support for science activities or even new forms of science production, e.g. "civic science"). Also interesting is the increasing engagement of the older scientific societies in such activities, in what can be seen not just as a response to the fact that their traditional role (namely communication among scientists through national scientific journals and conferences) is being increasingly threatened by the internationalisation of science, but also as evidence of the same dynamics that are supporting the emergence of new science dissemination organisations.

Nevertheless, a distinction was found among these two types of association, not just regarding the frequency of their public events, but also concerning the ways in which they are approaching the public. Most disciplinary scientific societies tend to opt for one-way, top-down formats, inspired by the "deficit model" of public communication of science (Trench 2008). They typically provide lectures, even if some are already becoming more aware of the need to find new forms/spaces for talking to the public, such as science cafes, and to collaborate with schools through science learning-by-doing activities.

Science dissemination associations tend to adopt more diversified and innovative formats. They often search for informal settings and for a deeper articulation between science and popular culture; they tend to promote more participative and "hands-on" experiences; many are trying to become directly engaged in local development initiatives; and some also support new "citizen science" projects. These approaches are likely to be bolstered by the fact that many of these associations have scientists, teachers and members of the general public among their members. In a sense, they may be closer to the spirit of the early days of scientific societies.

Bibliography

Bauer, M., N. Allum & S. Miller. 2007. "What can we learn from 25-years of PUS research? Liberating and widening the agenda". *Public Understanding of Science*, 16(1): 79-95.

Beck, U., A. Giddens, & S. Lash. 2000. *Modernização Reflexiva. Política, Tradição e Estética na Ordem Social Moderna*. Oeiras: Celta.

Ben-David, J. 1984. *The Scientist's Role in Society: A Comparative Study*. Chicago: University of Chicago Press.

Bensaude-Vincent, B. 2001. "A genealogy of the increasing gap between science and the public". *Public Understanding of Science*, 10(1): 99-113.

Bodmer, W. 2010. "Public Understanding of Science: The BA, the Royal Society and COPUS". *Notes and Records of the Royal Society*, 64(1): 151-161.

Bonney, R. *et al.* 2009. "Citizen science: a developing tool for expanding science knowledge and scientific literacy". *BioScience*, 59(11): 977-984.

Castells, M. 2009. *The rise of the network society: the information age: economy, society, and culture*. London: Wiley-Blackwell.

Chaline, J. 2002. "Les sociétés savantes en Allemagne, Italie et Royaume-Uni à la fin du XIXe siècle". *Histoire, économie et société*, 21(1): 87–96.

Clark, F. & L.I. Deborah. 2001. "Dimensions of civic science: introductory essay". *Science Communication*, 23(1): 5-27.

Conrad, C. C. & K. G. Hilchey. 2011. "A review of citizen science and community-based environmental monitoring". *Environmental Monitoring and Assessment*, 176(1-4): 273-291.

Costa, A. F., C. P. Conceição & P. Ávila. 2009. "Scientific culture and modes of relating to science" in *Knowledge and Society (Portugal in the European Context)* ed. A. F. Costa, F. L. Machado & P. Ávila, Lisbon, CIES-ISCTE-IUL/Celta: 61-84.

Dear, P. 1985. "Totius in verba: Rhetoric and Authority in the Early Royal Society". *Isis*, 76(2): 144-161.

Dierkes, M. & C. von Grote (eds). 2000. *Between Understanding and Trust. The Public, Science and Technology*, Amsterdam, Harwood Academic Publishers.

Evans, N. G. 2010. "Speak No Evil: scientists, responsibility, and the public understanding of science". *NanoEthics*, 4(3): 215–220.

Gascoigne, T. *et al.* 2010. "Is science communication its own field?". *Journal of Science Communication*, 9(3): 1-6.

Gonçalves, M. E. & P. Castro. 2002. "Science, culture and policy in Portugal: a triangle of changing relationships?" *Portuguese Journal of Social Sciences*, 1(3): 157–173.

Greenwood, J. D. 2007. "Citizens, science and bird conservation". *Journal of Ornithology*. 148(1): 77-124.

Gregory, J. & S. Miller. 1998. *Science in Public: Communication, Culture, And Credibility*. New York: Plenum Press.

Habermas, J. 1970. "Technology and Science as 'Ideology'". In *Toward a Rational Society; Student Protest, Science, and Politics*. Boston: Beacon Press.

Hartman, J. 1997. "The popularisation of science through citizen volunteers". *Public Understanding of Science*, 6(1): 69-86.

Hetland, P. 2011. "Science 2.0 : bridging science and the public". *Nordic Journal of Digital Literacy*, 6: 326-339.

Hin, L. T. W. & R Subramaniam. 1999. "How scientific societies can build better nations". *Science* 399: 633.

Irwin, A. & B. Wynne. 1996. *Misunderstanding Science? The Public Reconstruction of Science and Technology*. Cambridge: Cambridge University Press.

Jankowshi, N.W. 2007. "Exploring e-science". *Jornal of Computer-Mediated Communication*, 12(2): art. 10.

Leach, M., I. Scoones & B. Wynne. 2005. *Science and Citizens: Globalization and the Challenge of Engagement*. London: Zed Books.

Leydesdorff, L. & J. Ward. 2005. "Science shops: a kaleidoscope of science-society collaborations in Europe". *Public Understanding of Science*, 14, 4: 353-372.

Merton, Robert K. 1938. "Science, Technology and Society in Seventeenth Century England". *Osiris*, 4: 360–632.

Miller, S. 2001. "Public understanding of science at crossroads". *Public Understanding of Science*, 10(1): 115-120.

Miller, S. *et al.* 2002. *Report from the Expert group Benchmarking the Promotion of RTD culture and Public Understanding of Science*. Brussels: European Commission.

Raichvarg, D. 2005. *Sciences Pour Tous?* Paris: Gallimard.

Rasse, P. 2002. "La médiation scientifique et technique, entre vulgarisation et espace public". *Quaderni*, 46: 1-14.

Rilling, R. 1986. "The structure of the Society of German Chemists". *Social Studies of Science*, 16(2): 235-260.

Rogers, C. L. 1981. "Science Information for the Public: The Role of Scientific Societies". *Science, Technology & Human Values*. 6(36): 36–40.

Schofer, E. 2003b. "The Global Institutionalization of Geological Science, 1800 to 1990". *American Sociological Review*. 68(5): 730–759.

Shapin, S. 1996. *The scientific revolution*. Chicago: University Of Chicago Press.

Stehr, N. 1994. *Knowledge Societies*. London: Sage.

Teich, A. H. 2002. "AAAS and public policy: speaking softly and carrying a medium-sized stick". *Technology in Society*, 24(1-2): 167–178.

Trench, B. 2008. "Towards an analytical framework of science communication models", in *Communicating Science in Social Contexts. New models, new practices*, ed. C. Donghong *et al*: Springer, 119-135.

Zilsel, E. 2000. "The sociological roots of science". *Social Studies of Science*, 30(6): 935-949 (1942).

Yearley, S. 2008. "Environmental groups and other NGOs as communicators of science" In *Handbook of Public Communication of Science and Technology*, ed. M. Bucchi & B. Trench. London: Routledge.

CHAPTER TEN

CAVES WITH LIFE AND UNEXPECTED REMAINS: THE CONTRIBUTION OF AESDA TO SPELEOLOGICAL KNOWLEDGE IN PORTUGAL

FREDERICO TATÁ REGALA, RUI LUÍS, RUI MERGULHO AND PAULO PACHECO

Introduction

The Associação de Estudos Subterrâneos e Defesa do Ambiente (AESDA) is a non-profit and Non-Governmental Organisation (NGO), with scientific aspirations, which has been developing activity officially since September 1992. The contributions of this Association to knowledge and scientific research are devoted to the diverse matters of speleology. There were relevant findings related to cave biology, archaeology and palaeontology, in addition to research work, inventory, registration and exploration of subterranean sites, either of natural genesis or resulting from human activity, from the north to the south of Portugal. In addition to AESDA's editorial output, through the *Trogle* magazine, the work carried out under the aegis of this association, or related to it, has been published in diverse specialised scientific journals and disseminated in national and international events. This chapter addresses the curricular course of AESDA and, although focusing on the main examples of the accomplished research, which in many aspects represent contributions to scientific knowledge, it also approaches some sociological issues and the collective dynamics that characterised these twenty years of associative experience.

Speleology

Speleology gathers very different strands of activity, which inhibits its clear and rigorous definition. Since it is not a subject taught in universities or a sport in the strict sense of the term, speleology has been thriving in a very unique niche, oscillating between a mainly recreational occupation and science, with enthusiasts and supporters in both fields, sometimes standing for clearly antagonistic positions. In fact, while the etymological origin of the term "speleology" points out to knowledge and science, most of the related societies who define themselves as speleological play an eminently entertaining role, especially devoted to the areas of nature sports and adventure. It is not within the scope of this chapter to discuss the accurate concept of speleology, but it is important to refer AESDA's stance within this matter and its contribution in the scientific framework, also glancing over a few sociological issues.

In this context, AESDA has followed an autonomous and independent path, conceptually assuming speleology as an expression of knowledge or, in other words, a scientific field. However, in real grounds and in practice, AESDA's membership has been a genuine laboratory of socialisation, a platform for interactions between people from very different social upbringings and with diverse backgrounds, either from the perspective of academic qualifications or specific areas of expertise. Within this environment, AESDA is a convergence spot with unusual characteristics, where the researchers, students or workers find common goals to dedicate themselves. And this is done merely by having the satisfaction of achieving these objectives, without any monetary or material gain, but rather usually at their own expenses, with the essentially logistical support from the Association. Although AESDA is, as already stated, a ground for socialising and leisure activities, the scientific and technical components are taken with the necessary seriousness and conviction, and this has produced interesting results, as summarised in this chapter.

AESDA's foundation and first steps

Before AESDA was born, in the late 1980s and early 1990s the founding members were already promoting speleological activities, by then under the aegis of another speleological association, the Espeleo Clube de Torres Vedras (ECTV). Among the most active associates of this club, some diverging opinions eventually emerged concerning the importance or primacy of the scientific and technical projects in comparison with actions of recreational and cultural nature in a broader sense. In due course these questions became insoluble, compelling a group of cavers to stand out, cherishing well defined objectives and speleological ideas. They were able to gather the necessary conditions for the feasibility of the projects envisaged, which would take place within a new associative setting devoted to speleological activities. AESDA was officially founded in September 11, 1992, becoming therefore the legitimate entity that would allow achieving the aimed projects (DR III series, no. 255 of April 11, 1992).

The early years were characterised by great difficulties but also by important discoveries. Without official headquarters or collective equipment, AESDA was precariously maintained by the individual resources of its members. In fact, this new speleological association emerged in Torres Vedras, where another organisation of the same type was already well established (ECTV), making it difficult to obtain the necessary support from the local authorities and other potential funders. However, some important human acquirements readily improved the associative body and skills, and the results soon emerged. At this first stage and resuming the work started by the founders of AESDA when they were still assigned to ECTV, there were findings that, even after these last two decades, continue to bring surprises and provide irrefutably interesting scientific data.

Biological collection

Despite the young age of the associates involved in the work and the lack of specific academic qualifications, self-taught research and the contacts established with researchers from the Faculdade de Ciências da Universidade de Lisboa (FCUL) and the Centro de Zoologia do Instituto de Investigação Científica Tropical (CZ-IICT), especially with the entomologists Travassos Dias and Luis Mendes, endorsed the first research initiatives. At the dawn of AESDA, some very interesting biological collections were collected in several caves, mainly in Estremadura. It must be mentioned, in this first phase, the collection of bat ectoparasites (Regala and Mergulho 1995) and several other invertebrates, some of which would only be thoroughly known and studied many years later.

The discovery of a silverfish (insect) by then identified as belonging to the species *Coletinia mendesi* (Wygodzinsky, 1980), earned special consideration. From the Colaride cave (Cacém), it was delivered for study to Luis Mendes at CZ-IICT. Until then, this species had only been referenced in the Algarve and Cordoba (Mendes 1996). Later, during AESDA's activities, new specimens were collected in Ibn Ammar caves

(Lagoa), but those did not get to be studied in the belief that it would be the same species. Notably and thanks to the unceasing evolution of science, the specimens collected in Colaride are currently under taxonomic revision, following the recent discovery of a species new to science, in the Algarve (*Squamatinia algharbica,* Reboleira and Mendes 2012), based on specimens collected more recently by a biologist outside AESDA, Sofia Reboleira, from the Universities of Aveiro and La Laguna (Canary Islands), during surveys that, although not related to AESDA, had the collaboration of one of the authors (FTR) in the fieldwork (Reboleira *et al.* 2012).

Another particularly interesting case was a set of beetles collected in 1991 during the activities promoted by an AESDA working team (although before its official foundation), namely two of the authors (FTR and RM), in Almonda cave, Torres Novas. Some specimens were delivered to Artur Serrano from the Faculdade de Ciências da Universidade de Lisboa (FCUL), but they would be studied under the MSc thesis of the aforementioned biologist (Reboleira 2007; Reboleira *et al.* 2009), who received more specimens that were collected at the same time but remained deposited in the AESDA Bioespeleological Collection. The study of these specimens resulted in the discovery of another new species, which was named *Trechus lunai* (Reboleira and Serrano 2009). Until the conclusion of the study, only four individuals were referred to this species, two of which had been collected by AESDA, including the one that was used to define the holotype for the species description (Reboleira *et al.* 2009).

In AESDA the main promoters of biological collections eventually came to specialise themselves in other subjects and biology ended up on a secondary level within the activities, but the fact is that, given the history, members are aware of the matter and it is not unlikely that new discoveries of this kind may still come to light.

Archaeological discoveries

The on-going prospection fieldwork for the discovery of new caves has proven productive. There are several caves referenced by AESDA, some of which were completely unknown and others were doomed to oblivion. Of these, some have proved to be interesting from the archaeological viewpoint. The most notable case is the Algar do Bom Santo, Serra Montejunto (Alenquer). Discovered and unobstructed by AESDA in 1993, this cave has proved to be an important and significant archaeological deposit, containing a vast prehistoric necropolis, more precisely from the Neolithic (Regala 1995).

This is one of the largest caves known to date in the limestone massif of Montejunto and the archaeological remains were found within several rooms and galleries, reaching approximately 285 square metres of use as necropolis. The skeletons of about 120 individuals were reported at the surface levels, and it is likely that there are many more (Duarte 1998a; 1998b). Besides the osteological remains, there are several objects included in this remarkable archaeological burial site, comprising ceramic vessels, artefacts made of flaked flint, polished stone and bone, perforated shells and adornment beads. The use of the cave as a necropolis is chronologically attributed to the second and third quarters of the fourth millennium BC, according to the radiocarbon dates (Duarte 1998a; 1998b).

The discovery had a strong impact in the scientific community, compelling Instituto Português do Património Arquitectónico (IPPAR) (the Portuguese national institution responsible for archaeological heritage) to immediately mobilise the necessary resources for the protection and excavation of the site. Therefore a solid structure was promptly built for the protection of the entrance and a first long archaeological campaign was promoted in 1994, from June to November.

The news echoed through the media soon, with several articles in national newspapers and television broadcasts. Especially important was the production and exhibition of a documentary exclusively dedicated to the discovery and research of Algar do Bom Santo, entitled *Terra Humana* (Human land), authored by Paulo Costa under the aegis of Radio Televisão Portuguesa (RTP), in 1995.

The collaboration of AESDA's associates involved in the discovery and early exploration played a decisive role in the conservation of the archaeological remains that littered the cave. This is one of the rare cases, at the international level, where the incursions were careful enough to keep all the remains preserved and undamaged in their precise original location. In addition to the discovery and the care taken with the preservation of the site, AESDA participate in three archaeological field campaigns conducted between 1994 and 1997, initially promoted by the IPPAR and then by the Instituto Português de Arqueologia (IPA), funded by these and other entities. This collaboration involved archaeological excavation and recording tasks (drawing and photography) as well as the comprehensive topographic survey of the cavity, which was entirely in the hands of AESDA's members.

Despite the efforts to keep the site safe, the fact is that, unfortunately, in November 2000, the cave door was found disrupted and part of the archaeological remains from Algar do Bom Santo had been stolen. It was an irreparable loss for science and a sturdy blow to the willingness of AESDA's members to publicise their discoveries and deliver them into the responsibility of supervisory authorities. In November 2007, during a site visit by a team of archaeologists interested in promoting further research, it was found that a break-in had occurred again. The steel gate was open but no more signs of theft or vandalism were detected this time. Authorised by IPA, AESDA sealed the gate, fusing it to the surrounding frame.

Even with these drawbacks, the cave holds an immense supply of archaeological information. There is actually a project coordinated by the archaeologist Antonio Faustino Carvalho, from the Universidade do Algarve, with funding from the Fundação para a Ciência e Tecnologia (FCT), to develop and deepen the investigations on the site of Algar do Bom Santo.

The example of Algar do Bom Santo is paradigmatic of the relevance and usefulness of a willful association. In addition to having discovered the site, AESDA remained in connection with the research developments and recording works, as well as the protection and monitoring, managing sometimes to act more diligently than the responsible public services, but always in conjunction with them.

Another less significant archaeological work also deserves mention because it was conducted under the sole responsibility of the Association. This one concerns a couple of objects related to adornment discovered in 2000 during an obstruction removal in the Maceira karst system (Torres Vedras/Lourinhã). In a small cavity that might communicate with the main network of karst galleries, in order to verify this, some deposits were removed, uncovering fragments of human bones and two drilled and shaped wild boar canine teeth.

The obstruction clearing work was then suspended in order to safeguard other possible archaeological remains contained in the sediments and the study of the exhumed artefacts was endorsed. Despite the paucity of the material, which would not allow to draw much information, this new archaeological site was recorded, correlated with a necropolis context from late prehistory (Regala and Luis 2007), in addition to the elaboration of an analytical compilation of the archaeological knowledge concerning the caves in the vicinity (Regala and Gomes 2001).

Besides the examples above, AESDA has referenced other cavities with archaeological interest that await the opportune moment for a conscious and responsible disclosure, and that may be the subject for future research.

Palaeontology: a panther in Porto de Mós

One of the most interesting works promoted by AESDA relates to the identification of a large carnivore specimen extinguished from the European territory since prehistoric times. It is a set of bones that rested deep in a gallery (nearly 100m deep) of the Algar da Manga Larga, in Porto de Mós.

The remains were already known by previous explorers of the cave from other speleological associations since the 1970s, but those were reported by C. Thomas (1985, 1998-99) as corresponding to the skeleton of a dog. In January 2003, AESDA carried out a recognition of this Algar (the team was composed of Rui Luis, Sisenando Simões, Valter Bruno and Luis Oliveira), taking the opportunity to conduct a photographic record of the bones. Later analysis of these pictures revealed that it was not a dog but instead, a felid larger than the *Lynx*, linking the specimen under consideration with a species extinct from Europe, presumably a leopard, *Panthera pardus* (L. 1758).

During another mission to the site, biometric data was obtained, confirming the importance of the discovery. The collection and preservation of these paleontological remains were then carried out. The studies were upheld thanks to the collaboration established between one of the authors (FTR) and a palaeontologist outside the Association, João Luís Cardoso (Cardoso and Regala 2006; Regala and Cardoso 2007).

The collection, photographic record and the detailed topography with the location of these paleontological elements were carried out under the aegis of AESDA, as well as the scientific drawing of the bones and part of the investigation itself. Also a series of exploratory activities were arranged for the interpretation of the provenance of these remains, and the phenomena that caused them to reach the place where they were found.

The discovery has received interest from the media, being reported in several newspapers, in National Geographic Portugal (November 2011) and in television (news broadcasted by TVI, available online[1]). The interest caused by this research inspired an exhibition exclusively dedicated to it, at the invitation of the Porto de Mós Municipality, named "Fossil of the female leopard found on the Plateau de Santo António, Porto de Mós", which was held in that city from October 2009 to January 2010.

---

[1] *In* http://www.tvi24.iol.pt/videos/video/13137705/1475 (accessed in 2013/01/07).

The studies were widely disseminated within the scientific circles, being described in lectures and other events, including the 14th International Congress of Speleology, Athens-Kalamos (2005). Following the publication in *Trogle* magazine and in the prestigious specialty journal, *Comunicações Geológicas,* the leopard bones were offered to the historical Museu Geológico (Geological Museum), today under the Laboratório Nacional de Energia e Geologia (LNEG), in Lisbon, where they are on display and available for scientific research.

Subterranean exploration: the most notable breakthroughs

It is in the field of subterranean research and exploration that AESDA is actually more skillful to carry out work without having the need to resort to external entities or researchers, without excluding, however, fruitful partnerships and collaborations.

Prospection and obstruction clearing in caves and shafts led to several discoveries, as well as during the exploration of caves previously referenced by others. It would be too tedious to mention all the findings, therefore, in this chapter we only refer to those currently considered most relevant. In the region of Torres Vedras and neighbouring areas, inventories of cavities and numerous surveys were conducted, some of which were published and others are awaiting publication.

An interesting example is the karst complex of Maceira (Torres Vedras and Lourinhã), where a few limestone caves were known and considered to be “small cavities”. Following a cave topography course given by AESDA in 1999, it was noticed that several of these cavities were connected and led to an extensive network of intercommunicating underground galleries. The next step was an exhaustive topographic survey, which allowed the detection of other obstructed galleries that were cleared latter and explored for the first time (Regala and Luis 2002). The interest of the resulting work, which revealed the existence of approximately 980 meters of galleries, justified the communication presented at the 13th International Congress of Speleology in Brasília (2001). Currently AESDA has been developing similar work, now almost completed, in the largest underground complex of cavities known in southern Portuguese karsts, the so-called Ibn Ammar Caves, Lagoa (Algarve).

Further north, in the Massif de Sicó, during a joint activity of AESDA with Núcleo de Espeleologia de Condeixa (NEC) and Grupo Protecção Sicó (GPS), in June 2005, a new passage was found and cleared in the Abismo de Sicó (Sicó Abyss). This passage would allow going beyond the mark of 100-meter depth and reaching the deepest known level in a cave from this limestone massif, -107 metres deep (NEC and GPS 2007).

Also within partnership projects with other associations, it is relevant to note the recent subterranean and cave diving explorations in shafts in Alandroal (Alentejo), especially in the Algar das Morenas, which was explored to its maximum depth for the first time. This cave is an important natural underground water reservoir connected to the Estremoz-Cano aquifer, which supplies five municipalities: Sousel, Estremoz, Borba, Vila Viçosa and Alandroal. The cave, a large 78 meters deep vertical shaft, is flooded until approximately 47 meters from the bottom (measured in March 2012), although having a notable variation of the water levels as the result of rainfall and harvesting volumes. This cave was the subject of previous surveys, especially the dives in August 1983 conducted by J. Michel and J.-P. Thiry from the Speleo-Club de Dijon, which could not manage to reach the bottom, according to their report archived in the Alandroal City Hall.

A collaboration of AESDA with the scuba exploration group XploraSub, resumed the exploration and registration of this cavity since August 2008 and all data is currently being processed for publication.

When it comes to great depths in a karst system, the most important achievement of AESDA is quite recent. It is the deepest cave presently known in Portugal, the Algar PaLopes, named after AESDA*'s* associate Paulo Lopes, who located it, during a field prospection promoted by AESDA, on December 11, 2010.[2] The speleological potentials of this cave from Santo António Plateau remained ignored until the members of AESDA made contact with it and realised its actual interest. During a series of exploratory actions exclusively promoted by this association, depths never before attained in a Portuguese karst were reached. The cave that previously held the depth record was the Algar dos Alecrineiros, also in the same area, with -220 meters, discovered and explored in 2008 by a French team (Société Speleo Archéologique de Caussade). Still under survey, the Algar Palopes was explored to a depth of -268 meters. In this case, the great purpose of AESDA is not to break records but to reach a deep karst in the Estremenho massif, with all the information it has to offer. Again, this discovery was the subject of media interest, with news in

---

[2] Later we learned that the cave was already known by the caving team of the *Parque Natural da Serra de Aire e Candeeiros* (*PNSAC*) and was referenced in its inventory of caves since 1989. However, according to the information issued by ICNB in March 2012, only the first vertical pit would have been descended and there are no further records of exploration.

the national press and SIC television network, broadcasted in June 5, 2011 (Castelo 2011).

AESDA's fields of action are not restricted to karst environments and the genesis of natural cavities. Research has also been carried out in underground spaces of anthropogenic genesis, such as mines or crypts. The Roman gold mines in Valongo are also a good example of the work done by AESDA, right from the times of its foundation. The exploratory actions revealed more than 600 meters of galleries that were unknown or partly considered legendary. The topographic survey revealed several galleries in this complex, sometimes of monumental dimensions, with deep vertical wells that reach directly about -70 meters (Regala 1998). From the archaeological point of view, mention must be made to the discovery, inside the mine, of one type of Roman oil lamp, the Dressel 20/Ponsich IIb2, decorated with the figure of a winged Victory, attributable to the first century, whose study is properly published (Maricato 2001).

Edition

The edition of the work in AESDA has been, since the beginning, a main purpose vehemently advocated, in view of the apathy that characterises the Portuguese speleological scenery for the past two decades, a period of time characterised by the scarcity of publications on this matter. In this respect, the year of 1995 is of the utmost significance in the history of AESDA since it is when the first issue of *Trogle* was brought to the public. This was the much ambitioned vehicle for editing and divulging research produced by the members, thanks to the kindness and support of a company from Torres Vedras, the Sogratol Sociedade Gráfica Torreense. In fact, the reduced academic training and inexperience of the young authors by then did not make them back down from the imperative of publishing, preventing the work from becoming fruitless along with all the expended efforts. Following this first issue, others have been published, although with wide time gaps, given that the work develops according to the resources at hand and the availability of AESDA members. It is worth noticing that speleology is a subject unknown by many, with little impact or application for the wider public, making it less easy to obtain support. However, five issues of this magazine have already been edited and the sixth is in preparation, showing a positive evolution in the quality of the results and their presentation, as a result of increasing experience.

In these editions, AESDA has insisted in not to close upon itself, by granting other similar entities and external researchers the opportunity to publish their work in *Trogle*, if related to underground research (see Grupo Protecção Sicó and Núcleo de Espeleologia de Condeixa 2007; Mendes and Matos 2002; Rodrigues *et al.* 2001). This way, the magazine has accrued value, also earning prestige with inputs from outside the Association, becoming the sole Portuguese publication of speleology open to the whole national speleological community. The fourth and fifth issues of the magazine were fully published in bilingual mode (Portuguese and English) and distributed across borders, in many countries, making *Trogle* the main physical vector for publicising Portuguese speleology internationally.

Conclusions and brief sociological approach

As previously mentioned, AESDA has been a stage for human interaction and relationships in a context of great social and cultural heterogeneity at many different levels. Nonetheless, with diverse but very solid aims, the Association is an aggregating cluster of motivations that, when it comes to the scientific environment, has the great virtue of providing to basically anyone interested, an approach for research with practical results, coming actively in touch with the technical and methodological aspects involved, regardless of the curriculum and training.

Looking back at these two decades of membership in AESDA, one can take the pulse of the influence that this entity had in the lives of some of the associates and related persons, and it is valid to claim that it was crucial for the main structural options in their careers. There are associates who assume that their professional choices were made after the direct influence of what they have learned to do in AESDA and became enthusiastic in speleological activities, whether in technical areas such as photography and topography, or in scientific fields such as geology, archaeology, etc… In fact, the scientific disciplines, often hermetic and considered impervious, are here placed within the reach of everyone to participate or at least take an interactive contact. Thus, in addition to the actual contributions to science and knowledge, AESDA is a privileged interface between producers of science and general society, and a swivel for the discovery of individual vocations. Not being a professional research entity, AESDA does not pretentiously claim ambitions that are outside the scope of its modest means, but aims to nurture a scientific legacy of improved quality at every step, and ensures that all the effort of its members is not cast out to oblivion. For this reason, it is a prerogative of AESDA to take the imperative of publishing work with seriousness, even if not always with the same level of professionalism and quality, as this depends obviously on the means, availability and qualifications of the authors.

Although some expert studies have been mainly driven by entities outside the AESDA, the fact is that research conducted under the auspices of this Association brought data that remained totally or partially ignored by the lights of science. Its contribution is effective even if occasionally sidelined during the presentation of the results of some scientific discoveries. Such contributions are always the result of a genuine curiosity and interest, an unquenchable thirst to see beyond what is known, to be where nobody has been before, bringing to light the hidden secrets of knowledge in the realm of perpetual darkness... that is the powerful engine that moves AESDA's associates to search the underground spaces, a personal and collective investment at all levels. This explains what compels someone to defy the risk of wandering along unfamiliar places, often in inhospitable and uncomfortable conditions; this can only be thoroughly evaluated by those who go through that experience.

Acknowledgments

Authors are indebted to many of the AESDA's members and collaborators that are the human basis of all the achievements.

Bibliography

Cardoso, J. L. & F. T. Regala 2006. "O leopardo, *Panthera pardus*, L., 1758, do Algar da Manga Larga, Planalto de Santo António, Porto de Mós". *Comunicações Geológicas*, 93:119-144.

Castelo, C. 2011, "Um recorde no Interior da Terra", *Expresso*, 3 de Junho: 34.

Duarte, C. 1998a, "Escavações arqueológicas no Algar do Bom Santo", *Trogle*, 2: 9-15.

—. 1998b, "Necrópole neolítica do Algar do Bom Santo (contexto cronológico e espaço funerário)", *Revista Portuguesa de Arqueologia,* 1 (2): 107-118.

Grupo Protecção Sicó & Núcleo de Espeleologia de Condeixa 2007, "O Abismo de Sicó", *Trogle,* 5:24-29.

Maricato, C. 2001. "Breve notícia sobre a lucerna de Valongo", *Trogle,* 3: 14.

Mendes, L. F. 1996, "Further data on the Nicoletiidae (Zygentoma), with description of a new species from Mauritius", *Revue Suisse de Zoologie*, 103 (3): 749-756.

Mendes, J. R. & S. Matos 2002, "Gruta do Frade: o tesouro da Costa Azul, um monumento a preservar", *Trogle*, 4: 22-29.

Michel, J. & J.-P. Thiry 1983, *Relatório de exploração inédito do Spéleo-Club de Dijon*, Câmara Municipal do Alandroal archive.

Regala, F. T. 1995, "Montejunto revela necrópole neolítica de importância mundial", *Trogle*, 1: 14-15.

—. 1998, "Minas Romanas de Valongo", *Trogle,* 2: 3-8.

Regala, F. Tátá & J. L. Cardoso 2007, "Vestígios paleontológicos do Algar da Manga Larga", *Trogle*, 5: 4-17.

Regala, F. T. & E. H. Gomes 2001, "Grutas Arqueológicas da Maceira", *Trogle,* 4: 12-21.

Regala, F. Tátá & R. F. Luis 2007, "Os dentes de javali afeiçoados da Cova da Presa, em Ribamar, Lourinhã", *Trogle*, 5:18-23.

Regala, F. T. & R. Mergulho 1995, "Notícia da recolha de exemplares do *Ixodes (Eschatocephalus) vespertilionis* (Acarina, Ixodidae)", *Trogle,* 1:16-18.

Regala, R. & R. Luís 2002, "Grutas do Cabeço do Sino, estudo geomorfológico", *Trogle*, 4: 5-11.

Reboleira, A. S. 2007. *Coleópteros (Insecta, Coleoptera) cavernícolas do Maciço Calcário Estremenho: uma abordagem à sua biodiversidade.* MSc dissertation, Departamento de Biologia, Universidade de Aveiro, Portugal.

Reboleira, A. S., F. Gonçalves, P. Oromí & L. Mendes 2012, "*Squamatinia algharbica* gen. n. sp. n., a remarkable new Coletiniinae siverfish (Zygentoma: Nicoletiidae) from caves in Southern Portugal", *Zootaxa,* 3260:33-46.

Reboleira, A.S. P.S., F. Gonçalves & A. Serrano, 2009, "Two new species of cave dwelling *Trechus* Clairville, 1806 of the *fulvus*-group (Coleoptera, Carabidae, Trechinae) from Portugal", *Deutsche Entomologische Zeitschrift*, 56 (1), 101–107.

Rodrigues, L., A. Rainho & J. Palmeirim, 2001, "Os Morcegos Cavernícolas Portugueses", *Trogle,* 3:22-27.

Thomas, C. 1985, *Grottes et algares du Portugal*. Edição do autor, 230 p. Lisbon.

CHAPTER ELEVEN

GROUP OF STUDIES IN HUMAN EVOLUTION: ITS ROLE IN SOCIETY AND IN SCIENCE

SANDRA ASSIS AND VANESSA CAMPANACHO

Introduction

Scientists have an important role in developing interfaces between scientific knowledge, produced in universities and research centres, and the schools and/or general public. Despite the irreplaceable role played by schools in the production and distribution of knowledge, it is important to highlight the work of the media, science centres and museums, industry-education programmes, out-of-school programmes and community educational initiatives in promoting new avenues of non-formal and informal education (Honeyman 1998). According to Honeyman (1998: 1) non-formal education can help surpassing many hiatuses and scientific misunderstandings, increasing the opportunities for what he calls "scientific and technological literacy". Bell *et al.* (2009) further sustain that schools cannot act alone in the educational process, depending substantially on the initiatives of the civil society for the challenging task of teaching science.

Since 2005, a non-profit scientific association called Group of Studies in Human Evolution (GEEvH Grupo de Estudos em Evolução Humana) housed at the Department of Life Sciences of the University of Coimbra has developed a set of scientific and pedagogical activities in the field of human evolution and allied sciences. In addition to the diffusion of Darwin's theory in schools, museums and among the general public, the association aims to demystify the current picture that society has of the scientific milieu. It aims to show that: (i) science is a social enterprise and scientists do not work in a vacuum; (ii) science is embedded in the social and political context in which individuals live; (iii) scientific knowledge is tentative and experimental and can change (this is particularly true in the case of human evolution); and (iv) scientists are "normal" human beings and not "crazy doctors" locked in a laboratory (Bloom 2006).

In the following sections, an overview of the main issues that dictated the foundation of the GEEvH, its goals and achievements will be presented and described.

Evolutionism and the teaching of science: from the classroom to the public

"Darwin's legacy to modern science is unique and unquestionable" (Bracinha Vieira 2009, 9). This quote sums up the contribution that evolutionism brought to natural history, accomplishing the difficult task of explaining the natural world. Nevertheless, the concept of "evolution" was not completely absent from all the historical scenarios prior to Charles Darwin (Ridley 2004; Wood 2005; Jurmain *et al.* 2009). As pointed out by Kennedy *et al.* (2004), the ancient Greeks had already speculated about the origin of life, attributing the world's organic coherence to a gradual evolutionary process. Following the rise of Christianity in the Western World, the ideology that God created all living forms, with unchangeable proprieties, was generally accepted (Kennedy *et al.* 2004; Wood 2005; Cela-Conde and Ayala 2007). A true revolution happened on November 24, 1859, when the naturalist Charles Darwin published, in London, a book called *On the origin of species by means of natural selection or the preservation of favoured races in the struggle for life* (Mayr 2001). According to Darwin's theory, all organisms are the result of complex evolutionary processes that operate through the selection of the most well adapted forms in a process he termed "natural selection" (Mayr 2001; Kennedy *et al.* 2004; Mai *et al.* 2005; Ruse 2008). Almost at the same time, an identical model was published by Alfred Russel Wallace (1823-1913) (Jurmain *et al.* 2009). Both Darwin and Wallace regarded evolution as a ramified process that occurs from a common ancestry instead of a linear mechanism:

> ...in the actual state of nature it is almost impossible, the species being so numerous and the modifications of form and structure so varied, arising probably from the immense number of species which have served as antitype for the existing species, and thus produced a complicated branching of the lines of affinity, as intricate as the twigs of a gnarled oak or the vascular system of the human body (Wallace 1855, 189).

Charles Darwin also moved away from the anthropocentric views of his time, placing humans at the same level of the remaining living creatures (Avelar *et al.* 2004). In *The Descent of Man* Darwin argues that: "It is notorious that man is constructed on the same general type or model with other mammals" (Darwin 1871). According to Darwin, similar selective pressures were responsible for the whole biological diversity as well as for human origins: "In the future I see open fields for far more important researches. (...) Much light will be thrown on the origin of man and his history" (Darwin 1859, 414).

As a scientist, Charles Darwin coupled an extraordinary group of characteristics: he was an incessant observer of the natural phenomena, with a great theoretical capacity and critical analysis, as well as, persistence and rigor (Mayr 2001; Bracinha Vieira 2009). Furthermore, he was an interdisciplinary thinker bridging different branches of the natural sciences such as geology, botanic, zoology, palaeontology, among others (Bracinha Vieira 2009). Darwin built a solid tool that is still useful and updated presently. Some of his principles can be considered universal, namely: the relatedness of all organisms; the great diversity found in modern species; and the important role that evolution and natural selection played in the ever changing face of the planet's physical environment (Kennedy *et al.* 2004: 3). However, Darwin´s theory was not completely free of criticisms (Dennett 1995; Stefoff 1996; Rudolph and Stewart 1998; Jurmain *et al.* 2009; Mayr 2001; Ridley 2004). Many voices questioned Darwin's ideas, which were considered polemic or even heretic by the conservative and religious circles of his time (Dennett 1995). One of the major critiques of Darwin's theory lies on the difficulty of observing "evolution in action" in the human temporal scale (Mayr 2001). Two additional objections regard the lack of knowledge about hereditary rules, as well as the apparent gap between forms in nature (Ridley 2004). Some critics have also pointed out the non-testable proprieties of the evolutionary process, contrarily to other physical-chemical or biological phenomena (Mayr 2001). As stated by Kennedy *et al.* (2004), evolution is a process that operates within long time scales and it is dependent upon some unpredictable factors. Furthermore, its occurrence is impossible to isolate in a particular time or space, because it is a never-ending process that it is still on-going today. Only with the rediscovery of George Mendel's works about heredity, and with the discovery of DNA in the 1950s, were the mechanisms that control the intergenerational flux of traits clarified (Stefoff 1996; Kennedy *et al.* 2004; Ridley 2004; Ruse 2008).

Despite of all criticism and controversies involving Darwin's theory, it is clear that: (1) evolutionism allows the progress of science, constituting the theoretical basis of numerous sciences; (2) evolutionism plays an important role in human societies, since it explains many phenomena that affect our health and environment; and (3) evolutionism is a bridging theory that links different fields of research (Kennedy *et al.* 2004). Accordingly, the teaching of evolution is fundamental in scholarly learning processes (Almaça 1999). Jenkins (2009) states that natural sciences must be receptive to evolutionary theories since evolution is the key concept that allows us to understand the surrounding living world. For some researchers (i.e. Smith 2010; Yetişir and Kahyaoğlu 2010), the theory of evolution is regarded as an aggregating factor of scientific instruction. Furthermore, it has an important role in sciences such as medicine and epidemiology, for example in the study of the co-evolutionary process between hosts and pathogens, as well as in agriculture and in nature conservation (Kennedy *et al.* 2004; Jenkins 2009). In spite of the above-mentioned benefits, the theory of evolution seems currently forgotten in our educational system. Numerous factors can be pinpointed such as: (1) naive interpretations; (2) ideological and religious views or simple lack of knowledge; (3) old fashioned textbook approach and/or absence of updating concerning the recent scientific discoveries; and (4) the optional nature of this subject in science classes (Bracinha Vieira 2009; Smith 2010). Another important factor to consider is the lack of communication between scientific institutions and schools and/or the general public. With regard to the issue of human evolution, the difficulties enunciated are also present in the Portuguese educational system (for a detailed review see Assis and Carvalho 2012).

The Group of Studies in Human Evolution GEEvH: goals and achievements

To overcome the limitations faced by Portuguese instructors in teaching human evolution, a non-profit organisation called Group of Studies in Human Evolution (GEEvH Grupo de Estudos em Evolução Humana) (http://geevh.jimdo.com/) currently housed at the Department of Life Sciences of the University of Coimbra was founded in 2005 by a group of students attending the Master Course in Human Evolution (class of 2004-2006) in the former Department of Anthropology (University of Coimbra, Portugal). Based on a non-formal and informal approach to science, the main goals of the association are: (1) the scientific and pedagogical promotion of the multidisciplinary concept in Human Evolution studies; (2) the promotion and organisation of scientific meetings, congresses, conferences, courses, workshops, among other events; and (3) the interaction with other scientific institutions of national or international nature, focusing on Human Evolution knowledge. In summary, it aims to clarify misunderstandings that may populate common sense regarding the subject of human evolution and allied sciences.

In the next section, an overview of the main interdisciplinary activities held on a voluntary basis by the GEEvH members will be reported. It is important to highlight that all activities performed had in consideration the characteristics of the audience (i.e. age) and are divided into two types: the pedagogical and the scientific promotion of studies in human evolution.

Pedagogic promotion of human evolution interdisciplinary studies

Engagement with the school population is mainly achieved through pedagogic and interdisciplinary workshops. The workshops are part of the project "Brincar com a Grande Árvore da Evolução" (Playing with the Big Tree of Human Evolution, project VI: CV/PVI 1845) developed in the scope of the Ciência Viva VI programme and sponsored by it (Assis and Carvalho, 2012). Ciência Viva was created in 1996 and is affiliated to the Portuguese Ministry of Science (see Chapter Nine). This Agency aims to promote the scientific and technological knowledge among younger individuals and the general public, not only through "science for all" funding projects but also through the creation of a network of science centres distributed throughout the country. In addition to Ciência Viva, GEEvH has also been sponsored by the former Museu Antropológico – Museu de História Natural da Universidade de Coimbra for acquiring material supplies. Presently, GEEvH counts with eight distinct pedagogic workshops that have been presented mainly in schools, museums and hospitals. Developed on an informal and non-formal basis, their purpose is to disseminate the interdisciplinary concept of human evolution through pedagogical and experimental experiences directed at students of different age ranges (Assis and Carvalho 2012). The scientific areas covered are: life origin and biological diversity; human evolution; human osteology and bioarchaeology; forensic anthropology; primatology and prehistoric art (Assis and Carvalho, 2012). All workshops consist of a pedagogical model divided into two parts: concise theoretic exposition about a specific subject, followed by didactic and ludic games and/or hands-on activities (Assis and Carvalho 2012). In the interactive exercises, the topics learned are applied and, overall, students have shown an easy acquisition of the subjects taught (Assis and Carvalho 2012). An attempt to fit the discourse to the audience is always made by the workshop monitors; nevertheless, and as Assis and Carvalho (2012) point out, a re-formulation is needed with regard to special needs audiences. Currently, GEEvH is encouraging the creation of new workshops and provides training for monitors.

During the period of implementation of the project “Playing with the Big Tree of Human Evolution” (between 2005 and 2009), the workshops reached 1,678 individuals (Kindergarten: n= 265; Primary school: n= 304; Elementary school: n= 971; High school: n= 100; and adults with special needs and gifted children: n= 38) (Assis and Carvalho 2012). In 2008, the project “Playing with the Big Tree of Human Evolution” was presented by Susana Carvalho at the Leverhulme Centre for Human Evolutionary Studies, University of Cambridge (United Kingdom) in a paper entitled “Playing with the Big Tree of Human Evolution: interdisciplinary project”. In the same year, GEEvH participated in the Comics, Illustration and Animation Competition in Leiria (Portugal) with a workshop about prehistoric art, a collaboration with the illustrator Luis da Silva. In 2009, some pedagogical activities were also developed by Susana Carvalho at a school in Seringbara (Guinea Conakry).

The workshops were also presented to a countless number of participants in the following events in Portugal:

- 2005 to 2006: Exhibition Habitantes e Habitats (Exposition Inhabitants and Habitats) in the Municipality of Leiria
- 2008: Fórum Ciência Viva (Life Science Forum) in Lisbon
- 2008: Sobre as Pegadas do Passado, Museu Antropológico – Universidade de Coimbra (Upon the past footsteps, Anthropological Museum of the University of Coimbra) in Coimbra (1st edition: October-December 2008; 2nd edition: January-July 2009)
- 2008: Agreement with the Museu D. Diogo de Sousa (D. Diogo de Sousa museum) in Braga for the development of pedagogical activities between December 2008 and April 2009
- 2009: Feira da Ciência e Tecnologia de Alcobaça (Science and Technology Fair of Alcobaça) in Alcobaça
- 2009: IX Semana Cultural da Universidade de Coimbra, Museu da Ciência da Universidade de Coimbra (Ninth Cultural Week of the University of Coimbra, Museu of Science of the University of Coimbra) in Coimbra
- 2012: Noite dos esqueletos (Skeletons Night) in Figueira da Foz

The pedagogic workshops were mainly carried out in Portugal, in the following districts: Braga, Faro, Guarda, Viseu, Coimbra, Leiria and Lisbon.

In addition to the workshops, in 2012, GEEvH, along with CIAS and FCT (Fundation for Science and Technology in Portugal), sponsored the publication of a children's book by Sandra Assis entitled *Uma visita de estudo diferente: o universo da Antropologia e da Arqueología* (An unusual school trip: the universe of Anthropology and Archaeology). The book presents the opportunity of disclosing what Physical Anthropology and Archaeology consist of, using an accessible language and colourful illustrations sketched by Maria Nunes (Yamonine).

Promotion of scientific studies

GEEvH has also co-organised national and international scientific meetings, mainly with the purpose of allowing the promotion and discussion of scientific works, as well as the creation of new partnerships between researchers and university students, especially from Anthropology and Archaeology. Some of these events were also open to the general public, establishing a rare dissemination channel of concepts regarding human evolution for adults in Portugal. All the scientific meetings are listed below by year:

- 2006: Co-organisation of the Workshop Estudos Interdisciplinares em Evolução Humana (Interdisciplinary Studies in Human Evolution) in the 15th International Union of Prehistoric and Protohistoric Sciences Congress. It was held in Lisbon (Portugal). The group also co-edited the proceedings book published in 2009.
- 2008, 2010, 2012: Jornadas Portuguesas de Paleopatologia (Portuguese Meeting of Paleopathology), held at the University of Coimbra, and as a co-organisation with CIAS. This event aimed to report scientific papers about pathology cases in past populations around the world. The next congress is scheduled for 2014.
- 2009: Workshop in Musculoskeletal Stress Markers (MSM): limitations and achievements in the reconstruction of past activity patterns, held at the University of Coimbra. It was also a co-organisation with CIAS. 75 researchers from 21 countries were present in this workshop (Santos *et al.* 2011). More details about the workshop are available at: http://www.uc.pt/en/cia/events/msm.
- 2009, 2010: Mostra de Trabalhos Científicos (Scientific works display), at the University of Coimbra. It consisted of the display of posters previously presented in congress and workshops, aimed at university students and the general public. This event was co-organised by CIAS and IMAR Institute of Marine Research (2009)

and CIBIO Research Centre in Biodiversity and Genetic Resources (2010).

- 2010 to 2012: In this period the group co-organised eleven conference cycles entitled *Arqueologia e Antropologia...Territórios de Fronteira* (Archaeology and Anthropology...Border Territories). This is a partnership with NAP (Archaeology and Paleoecology Nucleus) and the National Museum of Archaeology, held in Lisbon (Portugal). Each conference cycle consisted in paper presentations of three national or international investigators from different areas of research. These events were open to researchers, university students and the general public.

GEEvH also co-supported the First Bioanthropological Meeting (I BAM) along with CIAS. I BAM was a congress meeting, organised by students, held in May 31 and June 1, 2013, at the University of Coimbra. It had the participation of researchers from around the world (more details available at: http://bioanthropmeeting.yolasite.com/).

Besides organising scientific events, GEEvH has also been present at scientific meetings, with the objective of advertising the group's existence and of disseminating scientific studies and data among researchers. GEEvH was present in the Jornadas de Investigação em Arqueologia (JIA), in 2011, at the University of Algarve (Portugal) and in the Colóquio Associações e Ciência, in 2012, in Lisbon (at the origin of this book).

2012 was also the year that GEEvH managed to achieve one of its original aims, the creation of a free online publication. The journal's name is *Cadernos do GEEvH*, and the editors are Francisco Curate and David Gonçalves. *Cadernos do GEEvH* already published a first volume with contribution of researchers from Anthropology and Archaeology.

GEEvH uses the World Wide Web to divulge scientific news about human evolution, events, scholarships and work opportunities, through a mailing list available only to members. A wide range of useful scientific information is also made available for the general public in the group website (http://geevh.jimdo.com/) and its Facebook page.

Final considerations

In spite of its voluntary basis, GEEvH has met the purposes established since its foundation. By investing in the human evolution area, GEEvH has created, and will continue to create, new dissemination routes and opportunities that did not exist previously in Portugal; and thus establishing bridges between two distinct universes, academic research, mainly in Anthropology and in Archaeology, and the general public, especially the students in non-tertiary education. On the other hand, the academic collaboration between distinct scientific institutions allows a more straightforward access to national and international research, and the establishment of new scientific partnerships. Students are conveyed the latest discoveries, aiming to instill in them the interest in human evolution and allied sciences.

Acknowledgements

Our thanks go to the members of Grupo de Estudo em Evolução Humana, with a special thanks to the Office Board, to Dra. Maria Arminda Miranda, and to the Research Centre for Anthropology and Health. We are also thankful to Ana Delicado for the invitation to write this chapter. The second author has a PhD grant from Fundação para a Ciência e a Tecnologia (grant reference: SFRH/BD/77962/2011).

References

Almaça, C. 1999. *O Darwinismo na Universidade portuguesa (1865-1890)*. Lisbon: Museu Bocage, Museu Nacional de História Natural.

Assis, S. & S. Carvalho. 2012. "Evolutionism and the Teaching of Science: How Portugal Has Been "Playing with the Big Tree of Evolution". *Evolution Education Outreach* 5 : 445–452.

Avelar, T., M. Matos & C. Rego. 2004. *Quem tem medo de Charles Darwin?* Lisbon: Relógio d' Água.

Bell, P., B. Lewenstein, A. Shouse, & M. Feder. 2009. *Learning science in informal environments: people, places and pursuits*. Washington: National Academies Press.

Bloom, J. 2006. *Creating a classroom community of young scientists*. New York: Routledge, Taylor & Francis Group.

Bracinha Vieira, A. 2009. *A evolução do Darwinismo*. Lisbon: Edições Fim de Século.

Cela-Conde, C., & F. Ayala. 2007. *Human Evolution: trails from the past*. Oxford: Oxford University Press.

Darwin, C. 1859. *On the origin of species by means of natural selection, or, the preservation of favoured races in the struggle for life*. London: John Murray.

—.1871. *The descent of Man and selection in relation to sex*. London: John Murray.

Honeyman, B. 1998. "Non-formal and formal learning interactions: new directions for scientific and technological literacy". *Connect UNESCO International Science, Technology & Environmental Education Newsletter* 23, nº 1: 1-16.

Dennett, D. C. 1995. *Darwin's dangerous idea: evolution and the meanings of life*. London: Penguin Books Ltd.

Jenkins, K. 2009. "Evolution in biology education: sparking imaginations and supporting learning". *Evolution: Education and Outreach* 2 : 347-348.

Jurmain, R., L. Kilgore & W. Trevathan. 2009. *Essentials of Physical Anthropology*. Belmont: Wadsworth Cengage Learning.

Kennedy, D., B. Alberts, D. Ezell, T. Goldsmith, R. Hazen, N. Lederman, J. McInerney, J. Moore, E. Scott, M. Singer, M Smith, M. Suite, & R. Wood. 2004. *Teaching about evolution and the nature of science.* Washington: National Academic Press.

Mayr, E. 2001. *What evolution is*. London: Phoenix Paperback.

Ridley, M. 2004. *Evolution*. Malden: Blackwell Science Ltd.

Rudolph, J. & J. Stewart. 1998. "Evolution and the nature of science: on the historical discord and its implications for education". *Journal of Research in Science Teaching* 35, nº 10: 1069-1089.

Ruse, M. 2008. *Charles Darwin*. Malden: Blackwell Publishing Ltd.

Santos, A. L., F. Alves-Cardoso, S. Assis, & S. Villote. 2011. "The Coimbra Workshop in Musculoskeletal Stress Markers (MSM): an annotated review." *Antropologia Portuguesa* 28: 135-161.

Smith, M. 2010. "Current status of research in teaching and learning evolution: I. Philisophical/Epistemological issues". *Science & Education* 19, nº 6-8: 523-538.

Stefoff, R. 1996. *Charles Darwin and the evolution revolution*. Oxford: Oxford University Press.

Wallace, A. R. 1855. "On the law which has regulated the introduction of new species". *Annals and Magazine of Natural History* 16: 184-196.

Wood, B. 2005. *Human Evolution: a very short introduction*. Oxford: Oxford University Press.

Yetişir, M. İ. & M. Kahyaoğlu. 2010. "Pre-service teachers' attitudes towards teaching of evolution theory". *Procedia Social and Behavioral Sciences*, 2: 1720-2724.

CHAPTER TWELVE

SCIENTISTS, ENVIRONMENTALISTS AND THE NUCLEAR DEBATE: INDIVIDUAL ACTIVISM AND COLLECTIVE ACTION

ANA DELICADO

Introduction

Science has long ago come down from the "ivory tower" and became engaged in social issues: issuing risk assessments and expert advice, promoting the diffusion of innovations, taking sides in scientific and technological controversies, developing actions and initiatives aimed at society at large, establishing cooperation with non-governmental organisations. Though this is often due to the efforts of individual scientists or research organisations, socially-oriented collective action in academia is becoming more frequent.

Schoefer's work on international scientific societies registered "a major shift from science associations organised as an inward-looking profession to science associations organised around social problems and societal concerns, often sustaining advisory positions (or other authoritative roles) in relation to the United Nations or particular national governments" (2003: 81). The author labels these latter organisations as "socially oriented", defining as their activities: "1) bringing scientific information to the citizenry or policy makers (...), 2) promotion of science or science policy that directly ameliorates social problems (...), 3) promotion of ethics in the application of science" (2003: 85). Their growth is linked to the increasing importance conferred to science as a "dominant model for organising social activity" (2003: 97) and to the integration of scientific expertise in decision-making. The environment, energy, development, peace, health, education, and public understanding of science are just a few of the areas in which these new associations intervene (see Chapter Eight in this book). Also, traditional disciplinary scientific societies are increasingly drawn to social debates and requested to provide policy advice and to participate in governance, both on matters of science policy and on issues with a techno-scientific dimension (such as health, environment, or risk). Scott *et al.* (2008) highlight the substantial increase in the engagement of scientific societies in policy matters, specifically in the field of natural resources in the United States. The authors signal the high number of associations that have created public affairs or policy offices and programmes, hired specialised personnel for these issues, organised seminars and issued statements, white papers, policy briefs and recommendations. They go on to defend that the interventions of associations are more effective than those of individual scientists.

Nuclear issues, both nuclear arms and nuclear energy, have been at the heart of a few scientific associations. The Pugwash movement was founded in 1957 as a result of the Russel-Einstein Manifesto against nuclear war. It was an advocacy group formed by scientists that "addressed governments on subjects like biological warfare, strategic arms limitation, and misuse of civilian nuclear technology" (Greenaway 2006: 178). This movement includes the Federation of German Scientists, created in 1959, as a result of the Declaration of Gottingen, in which nuclear scientists expressed their position against nuclear armament and vowed not to participate in military research. One of the key activities of this association is to award an annual prize to "whistle-blower" scientists, who denounce cases that endanger individuals, society, democracy, environment or peace. The Union of Concerned Scientists was created in 1969 in the US as a public interest organisation and one of its first battles concerned the safety of nuclear reactors (Dawney 1988; Rucht 1997). The International Council of Scientific Unions created in the late 1960s has a Scientific Committee on Problems of the Environment (SCOPE), which in turn originated the Committee for the Assessment of the Environmental Consequences of Nuclear War in 1983, whose report coined the phrase "nuclear winter" and was instrumental in the United Nations resolution concerning the reduction of nuclear arms (Greenaway 2006: 178).

This chapter looks into the participation of scientists in the debate concerning nuclear energy in Portugal in the last few decades, seeking to ascertain the role of individual and collective action. It will explore the alliances made between scientists, political actors, NGOs and citizens, the tensions within the scientific community and the impacts of "scientific" activism. It is based on document analysis (books, newspaper articles, reports) and interviews with the main actors in the events described below (scientists, environmentalists, activists from a local movement).[1]

Science, policy and governance in Portugal

A recent history of living under an authoritarian regime and the late development of science in Portugal may explain the Portuguese peculiarities in the role of scientific advice in policy making. In what regards science policy, centralised decision-making has been predominant. An advisory council (Higher Council of Science and Technology) was created firstly in 1979 (though it was never convened) and reactivated in 1986. However, its composition was unbalanced (more representatives from state laboratories than from universities) and the meetings were rare (Ruivo 1998, 257; Gonçalves 1993, 147), so it is safe to assume that its influence over policy was weak. The Council was restructured in 1996 and in 2003 was renamed the Higher Council of Science, Technology and Innovation, but short of publishing a report in 2004, no other activities are known. In 2006 the Council was renamed again (Coordinating Council of Science and Technology), but no legislation regarding its attributions or composition was ever published, so it remained inactive. 2011 brought a new change of name (National Science and Technology Council) and new members. The Council met regularly throughout 2012 and issued six advisory statements, concerning the Portuguese participation in international organisations, science funding programmes, scientific careers and the state laboratories. Nevertheless, there is no assessment of its actual impact on science policy.

The first framework law on scientific research in Portugal (1988) established that the scientific community ought to be consulted on matters of R&D investment (Gonçalves 1996, 55), but in practice this was restricted to the activities of the Higher Council and participation was

[1] This chapter draws on research carried out under several projects funded by the Portuguese Foundation for Science and Technology: *SOCSCI Scientific Societies in Contemporary Science* (PTDC/CS-ECS/101592/2008), hosted by the Institute of Social Sciences of the University of Lisbon, in collaboration with CIES-IUL and SOCIUS-ISEG; *Nuclear Portugal: Physics, Technology, Medicine and Environment (1910-2010)* (PTDC/HC0063/2009), hosted by the Institute of Social Sciences of the University of Lisbon, in collaboration with CES, University of Coimbra, coordinated by Tiago Saraiva; *Sociotechnical Consensus and Controversies regarding Renewable Energies* (PTDC/CS-ECS/118877), hosted by the Institute of Social Sciences of the University of Lisbon, in collaboration with the University of Aveiro and the Centre for Research in Anthropology.

"merely symbolic" (Gonçalves 1996, 59). In 1998 a White Book on the Portuguese Scientific and Technological Development was compiled, drawing on documents prepared by the Ministry but also on meetings with the scientific community throughout the country and contributions in an online forum. However, there is no information on how (and if) the White Book had any influence over science policy.

As to other fields with a scientific or technological dimension (environment, health, risk), even though in recent years public authorities and private companies have increasingly resorted to scientific advice, according to Gonçalves (2002) there is a lack of institutionalised forms of doing so, which tends to leave in the hands of actors the decision of when and how (which scientific disciplines and institutions to appoint) request the assessment of experts. Such is the case, for instance, of Environmental Impact Assessments:

> The relative weakness of the Portuguese scientific system and the lack of institutionalised forms of scientific advice for public administration have limited the scope for the preliminary discussion of the grounds for decisions taken within the context of the EIA procedures. This state of affairs, together with an inactive civil society, has contributed to maintaining the status quo of traditional administrative practice, which is most typically centralised, hierarchised and secretive. (Gonçalves, 2002: 250).

Recent risk cases examined by social scientists have shown that Governments, mainly motivated by the need to legitimise policy decisions, have zigzagged between different modalities for requesting scientific advice: sometimes to "independent commissions" made up of academics from reputed institutions, other times to their own experts working in state laboratories. Many of these processes have been riddled with controversy, both within the scientific community and with stakeholders (Gonçalves and Delicado 2009).

As to the participation of associations in science policy, in 1986 an association was formed, ADCT (Association for the Development of Science and Technology) whose main aim was to lobby in favour of the scientific community and it was represented at the Higher Council for Science and Technology (Gonçalves 1996; Ruivo 1998, 260). However, ADCT was disbanded in 1995. In the following year, the restructuring of the Higher Council foresaw the participation of five representatives of scientific societies, but since it had very little activity this probably never came into fruition; the following reorganisation of the Council in 2003 saw the disappearance of the representation of scientific societies. The

Federation of Portuguese Scientific Societies and Associations, created in 1991, also sought to have a place in the Council, but failed in its pretensions (Gonçalves 1996: 59).

Some scientific associations are included in advisory councils in other ministries, such as Health and Education (see Chapter Seven in this book). Other lobby government bodies and Parliament, pushing for policy reforms in their areas of expertise. For instance, in the last decade, the Association of Seismic Engineering held multiple meetings with government officials and members of parliament, issued statements and wrote letters aiming to change the laws governing the prevention and response to seismic risk (Delicado *et al.* 2013: 180).

Introducing nuclear energy in Portugal

Much like in other European countries (Joppke 1992; Lagaaij and Verbong 1999; Hecht 2009; Anshelm 2009), as well as in South America (Torres and Rudig 1983; Fernández 2011), plans for building a nuclear power plant in Portugal started to be drawn in the middle of the twentieth century, still under an authoritarian regime. Although nuclear research started at the University of Lisbon in the 1930s, it was only in the mid-1950s that the Government created both the JEN (Nuclear Energy Agency) and the Nuclear Energy Studies Commission in order to promote the "stable development of the national nuclear capacity" (Jorge and Costa 2001, 39). They were aimed both at research and application, namely the training of personnel for future nuclear facilities (Gaspar 2011). As a result, hundreds of scholarships were awarded to Portuguese students in nuclear physics, including for PhD studies abroad. JEN's initial attributions were the management of the uranium mines (see Chapter Five), but at the end of the decade a laboratory affiliated to JEN was created to carry out nuclear research, the Physics and Nuclear Engineering Laboratory (LFEN), heavily funded by bilateral agreements with the US under the "Atoms for Peace" Programme. The Portuguese Research Reactor was commissioned in 1959 and started functioning in 1961 (Jorge and Costa 2001; Gaspar 2011). However, according to Jorge and Costa (2001), LFEN always showed little interest in being involved in industrial applications of nuclear energy, preferring instead to focus on research and training.

In the private sector, a short-lived company, the National Company of Nuclear Industry (1958-1964), produced several technical studies and reports on the installation of a nuclear power plant in Portugal. Counting among its shareholders several electricity companies, industrial companies, a shipping company and a bank, the company's aim was to become "a qualified go-between in a future programme for building nuclear power plants in Portugal" (Jorge and Costa 2001, 72), but attempts to liaise with LFEN were thwarted. The main argument in the 1950s and 1960s favouring nuclear energy in Portugal was the existence of uranium mines, whose product could be used as nuclear fuel (see Chapter Five), and an expectation of growth of the electricity consumption that could not be met with existing hydroelectric resources. At the end of the following decade (late 1960s and early 1970s), the Government created a General Directorate for Industrial Nuclear Fuel and Reactors within JEN, in charge of licensing and supervising future nuclear power plants.

The revolution in 1974 and the ensuing political instability pushed back government plans for building the nuclear power plant, but only for a few years. The international context, mainly the oil crisis in the early 1970s, made it more urgent to assess the nuclear option. The electricity companies, later merged into a single company (EDP), had already been compiling technical reports in support of locating the first nuclear power plant in Ferrel, a small fishing and agricultural village near the coast, in the municipality of Peniche, 80 kilometres north of Lisbon. In November 1975 the nuclear option was discussed in a seminar on Energy Policy, in which the first critical voices were heard.

In March 1976 the government issued legislation indicating that the licence to build the nuclear power plant was going to be granted soon and the chosen location (Ferrel) was announced (where some preparatory building work had already begun). This event generated the public response described below. Faced with the public outcry, at the end of the year the government commissioned a White Book on the Nuclear Programme, an extensive technical report prepared by pro-nuclear scientists and engineers gathering contributions from several organisations. The White Book was finished in the following year and sent to the Prime-Minister, but it was only made public in 1980 and with a very limited diffusion (Pereira *et al.* 2010). According to its lead author, Jaime Oliveira, though the interest in resorting to nuclear energy was acknowledged, the White Book concluded "it would not be a mistake to delay setting up nuclear power plants in Portugal".[2]

Discussions continued throughout the final half of the decade and the issue was debated a few times in parliament, where the fierce opposition of some parties was expressed (Pereira *et al.* 2010). The National Energy Plans (1982 and 1984) foresaw the installation of between one and three

[2] Written statement sent after declining an interview with the author of this chapter.

nuclear power plants in the following decade, but these plans were not implemented. Paradoxically, in the same period JEN was discontinued and LFEN integrated in another state laboratory (LNETI National Laboratory of Industrial Engineering and Technology) concentrating all the laboratories that provided support to the industrial sector.[3] In the same period, the Office for Nuclear Safety and Protection was created under the tutelage of the Ministry of Industry, with the mission of controlling the safety of nuclear reactors and power plants, assessing and monitoring their functioning. In 1986 the Secretary of State for Environment officially announced the renouncement to the nuclear option.

Most commentators assess that the decision of not building the nuclear power plant in Ferrel was due not to public protest but rather to the lack of funds (Portugal had to receive support from the International Monetary Fund in 1981) and to the opposition of several ministers (Paiva 2008; Teixeira 2011). The final blow to the Ferrel project was a report from the Office of Nuclear Safety and Protection, assessing the seismological risk of the area and demonstrating that the proposed location did not meet the safety thresholds recommended by the International Agency for Atomic Energy (Ribeiro, Barriga and Cabral 2008).[4]

Ferrel: the first environmental social movement

Decades of an authoritarian regime in Portugal severely restricted the creation of environmental organisations and social movements in general. However, right after the 1974 Revolution the climate was ripe for collective action. Hundreds of local associations sprung up, such as residents' commissions (Pinto 2008):

> The revolutionary period (1974–75) witnessed a range of fascinating experiences of active citizenship and of attempting to bring together the two traditions of representative and participatory democracy. As the process unfolded (…) a plethora of social movements and citizen initiatives gave rise to the invention of new forms of participatory democracy. (Santos *et al.* 2004: 11)

---

[3] In 1994 LFEN regained its autonomy as ITN Nuclear Technology Laboratory, but it lost its state laboratory status in 2012 and was integrated in higher education, as part of the most prestigious engineering school in Lisbon.

[4] Curiously, right in front of the place where the nuclear plant was to be built, there is now a wave energy experimental facility. The current mayor, one of the leaders of the anti-nuclear protest, takes great pride in its municipality's renewable energy credentials.

Environmental NGOs (ENGOs) also only started to be formalised after 1974, but their number grew substantially over the following decades (Soromenho-Marques 2005; Figueiredo *et al.* 2001; Rodrigues 1996; Kousis *et al.* 2008). Figueiredo *et al.* (2001) also traced the surge in grassroots environmental protests between 1974 and 1994, noticing that right after the Revolution there was a substantial number of cases in which local populations rose against environmental threats, followed by a period of less conflict in the 1980s, only to rise again in the 1990s. These protests were related more frequently to waste disposal and treatment, followed by construction and infrastructure, manufacturing activities, wildlife areas and only in 9% of the cases with energy facilities.

Ferrel is widely considered as the first grassroots environmental movement, often celebrated as a founding moment of environmentalism in Portugal (Schmidt 2008). However, according to its protagonists, it started out a protest of the local population that only later was joined by environmentalists.

Poetical accounts of what happened in the morning of March 15, 1976, abound: "in response to the relentless chime of the church bells, the inhabitants of Ferrel went out into the street, stopped all the works, closed the ditches that had been opened and warned that if the works restarted they would come back to destroy whatever had been done" (APEAT 1981). According to one of the leaders of the local movement, the decision to march to the construction site of the nuclear power plant (in fact, just a meteorological station) was taken on the eve of that day, at a local meeting, spurred by the news on nuclear risks published in a local newspaper. The main reason for rejecting the nuclear power station was its environmental impacts, especially the threat to local fish species posed by warmer waters from the reactor. This was before Three Mile Island and Chernobyl, at a time that nuclear accidents were almost unheard of.

A local organisation was then formed, CALCAN Support Commission for the Struggle Against the Nuclear Threat, that sent a protest telegram to the Prime-Minister, signed by local associations, local newspapers, residents' committees, local business, trade unions and schools. Over the following months several embryonic ENGOs and environmental activists (Afonso Cautela and José Carlos Marques were leading figures) showed their support, local meetings were promoted with the participation of scientists (see below), and the protest continued. In February 1977, the environmental group "Living is Needed" launched a national appeal under the heading "We are all inhabitants of Ferrel", against the pro-nuclear policy of the government. A few months later, in June, a forum was held in a nearby city, Caldas da Rainha, from which emerged two statements,

"Portuguese Nuclear Programme Moratorium" and "People's Creative Intervention in Defence of a Non Depleted Environment". A folk singer, Fausto, wrote a song in which the nuclear power plant was mentioned.

In January 1978 the last major event of this protest took place. The "For Life and Against Nuclear" Festival was held at Caldas da Rainha, bringing together close to 3,000 participants, among which several scientists and artists, as well as environmental activists and local population. Debates, concerts, workshops and other activities were promoted, alongside a peaceful demonstration that again reached the location in which the nuclear facility was meant to be built: in a symbolic gesture of reclaiming that land for agriculture, the demonstrators planted potatoes.

In the late 1970s other demonstrations and actions were organised against the nuclear option in other parts of the country,[5] also strongly influenced by the "anti-nuclear wave that swept across advanced industrial democratic nations by the late 1970s" (Gil Nave 2001, 374; see also Touraine *et al.* 1980, Kitschelt 1986; Rucht 1997), but the movement fizzled out when the plans for the nuclear power plant were abandoned. Since then, many of the ENGOs involved in these events have ceased to exist but others, much stronger, have taken their place.

Ferrel has remained for many the blueprint for other grassroots environmental movements. From the construction of a thermal power plant in Vila Nova de Anhã (Kousis 2004) to the co-incineration of industrial waste (Matias 2004; Gonçalves and Delicado 2009), from mobile phone masts to toxic discharges in watercourses, many technological hazards have been the target of public protest in Portugal in the past decades. In these cases, the public has come together in impromptu (or sometimes formalised) groups, has gathered scientific information (often if not always from foreign sources), has brandished social and technical arguments with more powerful opponents, has deployed all tools of political protest and even resorted to the terribly slow and expensive Portuguese judicial system to defend the health and safety of populations.

The role of scientists

Controversies around nuclear energy have a strong scientific nature and it is hardly surprising that scientists became heavily involved in the

[5] These sessions usually brought together scientists, environmentalist but also representatives from the fishing trade unions and other local stakeholders from Ferrel.

nuclear debate, both in Portugal and in other countries.[6]

On the pro-nuclear side, the government stance was supported by several technical reports (including the White Paper) produced by engineers and researchers working either for the electricity company EDP or for JEN/LFEN (on the lack of public trust in scientific data provided by private companies, see Covello and Peters 1997, Wynne 2002). They intervened in the public debate by making presentations at seminars organised by EDP (there was one in 1976 aimed exclusively at journalists), by giving interviews and writing opinion articles in newspapers and by publishing books. One of the leading figures in this group was Jaime Oliveira, coordinator of the team in charge of compilling the White Book, a PhD in Nuclear Physics and researcher at LFEN. He wrote several books aimed at the general public (Oliveira 1977; Oliveira and Martinho 1977) with the purpose of

> contributing to the public debate of such an important (and exciting issue) such as the peaceful use of nuclear energy in Portugal (...) a global vision that will make easier [for the reader], I hope, to appreciate its relative importance. Against the emotional and so often ill-founded relations they raise, I seek to counterpoint the serenity that stems from the scrupulous respect for scientific objectivity (Oliveira 1977: 6).

The main arguments put forward by scientists defending the nuclear option in the 1970s were that it was the sole solution for solving the energy crisis and sustaining the predictable rise in electricity consumption, that uranium mining resources could be put to use and that environmental risks were manageable: the probability of risks was very low, safety standards were high, other facilities of energy production were more polluting and hazardous.

However, not all the scientists at LFEN were in favour of the nuclear power plant. Some publicly expressed scepticism towards the plan, arguing that the existing financial and human resources were insufficient for operating such a major infrastructure.

Most academic scientists took the anti-nuclear stance and sided with the local populations. One of the most vocal was Delgado Domingos, professor at the most prestigious engineering school in the country (IST)

---

[6] In 1975, 4,000 French scientists and engineers signed a petition questioning nuclear power, although A. Tourraine and his team discovered that some of the activists were more concerned about the bureaucratic organisation of their laboratories and were "profoundly sceptical of expertise" (Weart 1988, 352; see also Touraine *et al.* 1980)

and an expert on energy, who started intervening in the early discussions in 1975 and is, to this day, a permanent fixture in all the debates on nuclear energy in Portugal. He started by writing articles in newspapers and publishing several books on the subject (Domingos 1978), but it was his intervention in the Ferrel case that is more outstanding. He was invited to address the local population at a public session, right after the "invasion" of the construction site:

> It was in a theatre, there were a lot of people there, I was very anguished, but I explained them some fundamental data, the issue of the currents. There is sheer personal credibility, but people didn't know me, I was sensitive to that. They were fishermen, so I remembered the example of an accident in which they drowned, so I said "you don't need experts, you know the currents, you know if someone drowns here where you can pick him up". It was immediate harmony, they raised me on their shoulders, I was very embarrassed. (Interview with Delgado Domingos, 2004)

He returned to Ferrel subsequently, to the Festival in 1978 and continued to speak publicly against the nuclear power plant. Some of the interviewees stated that he attended a session aimed at local residents organised by the promoter of the power plant in disguise and discombobulated the engineers by asking challenging technical questions.

Other scientists were also taking individual stances against the nuclear options, by writing opinion articles in newspapers and participating in seminars and events, as well as in the 1978 Festival in Caldas da Rainha. Although biologists, engineers and chemists also became involved in the debate, many of these researchers were nuclear physicists, trained in France, where the nuclear controversy was rife, bent on using their expert knowledge to enlighten the public and raise awareness of the nuclear risks:

> I was late for the revolution, I was very upset because I wasn't here on April 25, 1974 (…) I was mortified by missing all the important historical events and I decided to become involved, not exactly on politics, but on social issues. It seemed that my voice, as a scientist (I had already a PhD), as a physicist, I had something to say. My involvement in political life was to contribute to raise awareness in the population about what nuclear was and what was at stake, the risks. I must confess I was very much influenced by France, by the anti-nuclear movements in France. (Interview with Carlos Matos Ferreira, 2013)

As to the role of scientific associations in this controversy, the Physics Society might have been a key player, but the government made no formal request for advice nor had at the time any consulting body in which the

Society could take part. On the other hand, during this period the Physics Society experienced several upheavals that possibly prevented a formal intervention in this issue: the separation from the Chemistry Society in 1974, the absence of a head office until 1977, the interruption of the publication of its bulletin between 1974 and 1978. Nevertheless, most of the scientists that took part in the debate, both pro and anti-nuclear energy, were members of the Physics Society and even part of its governing bodies.

However, collective action does not reside solely in formal associations. In what might be considered as a "social movement of scientists" against the nuclear option, in June 1977 a group of 110 scientists and engineers (many from the electricity company) signed a Manifesto on Energy Policy, asking for a national debate on nuclear energy, which was published in the leading cultural magazine of that time, *Raíz e Utopia*. The main claim of the Manifesto was that there was insufficient scientific evidence to support the nuclear option and that such a major decision had to be preceded by a wide public debate. A year later, this group (which by then had grown to 200 members) formed a "Commission for Promoting the National Debate on the Nuclear Option" and held a press conference, asking for the White Paper to be made public and announcing the creation of several work groups for studying issues such as safety, the fuel cycle, economics, the participation of national industries, energy alternatives and the biological effects of radiations. The Commission was led by two university professors from the engineering school IST, both of them nuclear physics, Carlos Matos Ferreira and Tito Mendonça. Their main demand continued to be an independent scientific evaluation of nuclear energy in Portugal, but the group also put forward several arguments against nuclear: the economic competitiveness of nuclear energy in the long run was doubtful, ensuring safety standards would increase costs enormously, there were no environmental impact assessments done in Ferrel (concerning maritime currents, atmospheric conditions, ecology, fish resources, thermal and radioactive pollution).

In 1979, after the Three Mile Island accident, representatives from this Commission published an opinion article in a national newspaper in which they reaffirmed their position against a nuclear power plant, by arguing that accidents were a highly probable and serious risk and the lack of nuclear engineers in Portugal increased that risk. They also pointed out the threat of environmental contamination, the absence of a solution for nuclear waste, that hydrological resources could be better used and that uranium imports would increase the dependency of Portugal from abroad.

In November 1978, another group of scientists, who had been meeting regularly in energy seminars, published a book on this issue (Moura *et al.* 1978). The authors came from different backgrounds (economics, physics, and engineering) and worked at different institutions (LFEN, IST, the University of Lisbon, EDP) and the contents of the book are mainly technical. Its purpose is stated in the introduction:

> [the book] stems from the conscience that technicians and scientists have a civic mission of public enlightenment that must be fulfilled within their areas of expertise and to the best of their abilities (…). The authors aim to enlighten the reader about the problems of real importance that the use of nuclear energy raises in our country, in order for him to distinguish what is crucial from what is not and take a substantiated position (…). The debate about the nuclear power plant is an exemplary case of the participation that citizens could and should have in the discussion of undertakings and economic and social policy options that are determinant for the future of a region or country. The authors hope that this book contributes to helping the debate become more useful and fruitful (Moura *et al.* 1978, 16).

Nevertheless, it is the conclusion of the book that points towards the standpoint of the authors: they conclude that "on a national perspective, there is no technical nor economical basis for opting for a nuclear power plant" (Moura *et al.* 1978, 276), since it would not replace conventional energy sources, it would increase foreign dependency and there would be great difficulties in ensuring safety conditions in Portugal. There is also an undertone of denouncing the political and corporate interests that permeate the nuclear option.

Thus, in the nuclear controversy of the 1970s, scientists presented themselves as "enablers" of public participation, as a source of "enlightenment" to the general public, providing accurate and unbiased information to help society decide which side to take in the discussion. It cannot be but a reflection of the post-revolutionary climate this emphasis on public participation in major policy decisions. And though individual opinions and statements from scientists were still dominant in the debate, the scientific community also showed some ability to act collectively, which is not very usual in Portugal.

Gil Nave considers that in this case "what was at stake was the legitimacy of decision-making processes centred on the mechanisms of representative democracy and administrative technocracy". He explains the success of the anti-nuclear stance by a "weakly organised, though convincing, discourse-oriented action that strongly mobilised public opinion", which "took advantage of cleavages over the issue among state and governmental actors in one hand and parties on the other". However, the author also stresses the role played by "other powerful and highly influential actors (...) an active group of university experts on energy (...) influential leaders of conventional parties and of several governments of the period, and an elite of intellectuals and media opinion-makers" (2001: 347-348).

The issue emerges again in the first decade of the new century

Throughout the 1980s and 1990s the nuclear debate abated considerably. Although the issue was periodically raised whenever the national energy plans were under discussion or there was some problem in the Spanish nuclear power plants, it was only in the first decade of the twenty-first century that the discussion was reignited. In mid-2005, a group of businessmen started lobbying in favour of nuclear energy. In a press conference, they announced the intention of building a nuclear power plant and the creation of a company, Nuclear Energy of Portugal. They gradually gained the support of several semi-public companies, such as EDP, REN (electricity grid company), some former ministers and junior ministers, and the Governor of the Bank of Portugal. However, the government took no official position over the issue. Over the next few years, numerous conferences were held for debating this issue, bringing together foreign experts (from countries with nuclear energy), business representatives, pro and anti-nuclear scientists and ENGO representatives. A book was published in 2006 (Rodrigues and Azevedo 2006), gathering contributions both pro and anti-nuclear, again aiming to foster public debate. The issue was debated again in parliament. In 2010 a group of business representatives, engineers and scientists signed a manifesto for a new energy policy that criticised the option for renewable energies and claimed for the need to reassess other options, implicitly nuclear energy. The socialist government in power rejected these demands and renewed its commitment to invest in solar and wind power, as well as to build more dams. A new manifesto was issued in 2012, this time explicitly in favour of nuclear energy, but the Minister of Economy dismissed it promptly, saying that energy generation in Portugal was already exceeding power needs. Again, no consultation mechanisms in which scientific associations could take part were activated in any of these occasions.

During this period there were no grassroots movements against the nuclear option, mainly due to the fact that no locations for a power plant were ever openly discussed. However, in 2006, the 30th anniversary of the Ferrel movement was celebrated with a demonstration that brought together many of the scientists, journalists, environmentalists and local authorities involved in the events.

What has changed substantially in these three decades is the role of ENGOs in the debate. These organisations have become stronger (Soromenho Marques 2004; Figueiredo *et al.* 2001; Gil Nave 2001), with thousands of members; they receive public subsidies and are represented in advisory bodies (for instance, in the National Council for Environment and Sustainable Development CNADS). Many are headed by academics and develop a wide array of activities, including at times scientific research. They have taken the lead in the campaign against nuclear energy, issuing statements, organising press conferences and taking part in public debates. In 2006, a "Platform No to the Nuclear Option in Portugal" was formed, bringing together several ENGOs, citizens and other organisations.

Several scientists also manifested publicly their opposition to the nuclear option. Delgado Domingos remains one of the more sought after opponents of nuclear energy, invited to almost all seminars on this issue, but others have also followed suit, such as Manuel Collares Pereira (INETI), Eduardo Oliveira Fernandes (University of Porto), Jose Peças Lopes (University of Porto), António Sá da Costa (civil engineer, president of the Association for Renewable Energies), Frederico Carvalho (ITN). Other scientists have spoken in favour of the nuclear option, such as Carlos Varandas (head of the Research Centre on Nuclear Fusion, professor at IST), José Carvalho Soares (former head of ITN, professor at the University of Lisbon), José Luís Pinto de Sá (professor at IST), José Marques (ITN).

There is no clear division between scientific disciplines or institutions in this debate. The arguments remained fairly the same as in the 1970s, but the anti-nuclear camp has included the important role of renewable energies in Portugal in their argumentation. Conversely, the pro-nuclear camp argues that nuclear technology has evolved substantially in the past few decades, stepping up safety, and that breakthroughs in nuclear fission are imminent. This side has also taken advantage of the problem of climate change, arguing that nuclear energy is carbon-free (Jean-Baptiste and Ducroux 2003; Bickerstaff *et al.* 2008; Bang 2010; Pampel 2011).

In the 2000s, though individual activism is still predominant, the collective action of scientists has become much more significant. The Portuguese Physics Society followed the lead of its European counterpart, the European Physical Society, which in 2007 had issued a position paper in support of nuclear energy, defending the need "to inform the general public on how to assess relative risks rationally". The Portuguese association held a round table in the following year for discussing the position paper. The Society's president publicly sided with the pro-nuclear side and several articles were published in its bulletin advocating the need to resort to nuclear energy, although few explicit references to the Portuguese case were made. A few years previously, the Society had also sponsored an exhibition at the Science Museum of the University of Lisbon, designed by PANS Public Awareness in Nuclear Science Group of the Nuclear Physics European Collaboration Committee, which presented a very favourable portrait of nuclear research but also of nuclear energy (Delicado 2010).

The following year, a group of its members, most of them physicists, announced they were forming an association named ADN Association for the Dissemination of Nuclear Energy. Some of the scientists involved had taken part in the 1970s movement and had changed sides (João Caraça), others were new to the debate: Augusto Barroso (University of Lisbon), Carlos Fiolhais (University of Coimbra), Manuel Sobrinho Simões (University of Porto), Ana Arriaga (University of Lisbon), Ana Eiró (University of Lisbon), Clemente Pedro Nunes (IST), Fernando Carvalho Rodrigues (NATO), Fernando Parente (New University of Lisbon), José Dias Urbano (University of Coimbra), José Marques (ITN), José Sucena Paiva (IST), Paulo Santos (New University of Lisbon). A few came from other disciplines, such as medicine and electronics engineering. ADN was meant to be "a platform for debate and accurate information on nuclear energy" and was created to "increase the information and training on nuclear energy (...) [and] widen the debate to society in general", but the manifesto clearly defended the investment in nuclear energy and finished with the sentence: "all energy we consume on Earth comes from the sun and the energy from the sun is nuclear". However, this association was never officially created and, in the midst of a severe financial crisis, the issue of nuclear energy seems to have died down again in Portugal.

Final remarks

Nuclear energy is an exceptionally contentious techno-scientific issue and has been so for almost a hundred years. It is probably one of the most obvious subjects for examining the interplay between scientists and social movements. But it also seems a privileged standpoint for analysing the collective action of scientists themselves.

From this brief description of the Portuguese case, a few aspects should be highlighted. First, how the re-emergence of the debate 30 years after the initial outbreak brings back many of the same actors (though some have changed sides in the meantime) with similar arguments to justify pro and anti-nuclear positions. Second, how unlike other risk cases, the nuclear controversy motivated scientists not just to issue statements and advice but also to come together and act as a group and even, in more recent years, to form an association. And thirdly, how the role played by scientists cannot be seen in isolation either from the internal workings of the scientific field (the competition between scientific disciplines, the rivalries between institutions) or from the actions of other social actors (ENGO, business companies, the Government).

Concerning the specific issue of the role played by scientific societies, it is noticeable how much more vocal and organised they have become in the past three decades. Even though it seems that little margin is still being awarded to them in terms of policy advice and representation, the advantages of organised collective action seem to have gradually become recognised by the scientific community. Either by acting within existing associations or by forming new organisations, groups of scientists are striving to be heard and taken into account in important societal debates.

References

Anshelm, J. 2009. "Among Demons and Wizards: The Nuclear Energy Discourse in Sweden and the Re-Enchantment of the World." *Bulletin of Science, Technology & Society* 30 (1) (December 29): 43–53.

APEAT Associação Portuguesa de Ecologia Amigos da Terra. 1987. *Antes... durante... e depois de Chernobyl : o nuclear no mundo e em Portugal*, Lisbon: A.P.E.

Bang, G. 2010. "Energy Security and Climate Change Concerns: Triggers for Energy Policy Change in the United States?" *Energy Policy* 38 (4) (April): 1645–1653.

Bickerstaff, K., I. Lorenzoni, N.F. Pidgeon, W. Poortinga, & P. Simmons. 2008. "Reframing Nuclear Power in the UK Energy Debate: Nuclear Power, Climate Change Mitigation and Radioactive Waste." *Public Understanding of Science* 17 (2) (April 1): 145–169.

Cautela, A. *et al.*. (1977), *O suicídio nuclear português*, Lisbon: Socicultura.

Covello, V. T. & R. Peters. 1997. "The determinants of trust and credibility in environmental risk communication: an empirical study". *Risk Analysis*, 17(1): 43-54.

Delicado, A. *et al.* 2013. *Ciência, profissão e sociedade: as associações científicas em Portugal*. Lisbon: Imprensa de Ciências Sociais.

Delicado, A. 2010. "Exhibiting Science in Portugal: Practices and Representations in Museums" *Portuguese Journal of Social Science* 9 (1): 19-31.

Domingos, J. D. 1978. *Inteligência ou subserviência nacional?*, Porto: Afrontamento.

Downey, G. L. 1988. "Reproducing Cultural Identity in Negotiating Nuclear Power: The Union of Concerned Scientists and Emergency Core Cooling", *Social Studies of Science*, 18 (2): 231-264.

Fernández, J. R. 2011. "El Surgimiento de Las Comisiones de Energía Atómica En Argentina y Brasil (1945-1956)." *Eä Journal* 2 (3): 1–21.

Figueiredo E., T. Fidélis & R. Pires A. 2001. "Grassroots Environmental Action in Portugal (1974-1994)". In *Environmental Policies in Southern Europe, actors, institutions and discourses in an europeanizing society*, ed. K. Eder & M. Kousis, Kluwer Academic Publishers. Dordrecht: 129-221.

Gaspar, J. 2011. "O Reactor Português de Investigação na Encruzilhada com o Desenvolvimento da Física Moderna em Portugal e os Átomos para a Paz (1952-1961)." *Gazeta de Física* 34 (2): 58–62.

Gil Nave, J. 2001. "Non-governmental groups and the state". In *Environmental Policies in Southern Europe, actors, institutions and discourses in an europeanizing society*, ed. K. Eder & M. Kousis, Kluwer Academic Publishers. Dordrecht: 343-364.

Gonçalves, M. E. 1993. "Ciência, comunidade científica e democracia em Portugal". In *Comunidade científica e poder,* ed. M. E. Gonçalves. Lisbon: FPASC/Edições 70: 133-150.

—. 1996. Mitos e realidades da política científica portuguesa. *Revista Crítica de Ciências Sociais*, 46: 47-67.

—. 2002. "Implementation of EIA directives in Portugal: How changes in civic culture are challenging political and administrative practice," *Environmental Impact Assessment Review* 22: 249–269.

Gonçalves, M. E. & A. Delicado. 2009. "The Politics of Risk in Contemporary Portugal: Tensions in the Consolidation of Science-Policy Relations". *Science and Public Policy* 36 (3): 229-239.

Greenaway, F. 1996. *Science International: a history of the International Council of Scientific Unions*. Cambridge: Cambridge University Press.

Hecht, G. 2009. *The Radiance of France: Nuclear Power and National Identity After World War II*. Cambridge MA: MIT Press.

Jean-Baptiste, P. & R. Ducroux. 2003. "Energy Policy and Climate Change." *Energy Policy* 31 (2): 155–166.

Joppke, C. 1992. "Models of Statehood in the German Nuclear Energy Debate." *Comparative Political Studies* 25 (2) (July 1): 251–280.

Jorge, H. M. & C. J. Costa. 2001. *O Reactor Português de Investigação no panorama científico e tecnológico nacional, 1959-1999*, Lisbon: ITN e Sociedade Portuguesa de Física.

Kitschelt, Herbert P. 1986. "Political Opportunity Structures and Political Protest: Anti-nuclear Movements in Four Democracies." *British Journal of Political Science* 16: 57–85.

Kousis, M. 2004. "Marine and coastal issues in local environmental conflict: Greece, Spain and Portugal". In *Contesting the foreshore: tourism, society and politics on the coast*, ed. J. Boissevain & T. Selwyn, Amsterdam: Amsterdam University Press: 205-231.

Lagaaij, J. A. C. & G. P. J. Verbong. 1999. "Different Visions of Power The Introduction of Nuclear Power in the Netherlands 1955-1970." *Centaurus* 41 (1-2) (April): 37–63.

Matias, M. 2004. "Don't Treat us Like Dirt': The Fight Against the Co-incineration of Dangerous Industrial Waste in the Outskirts of Coimbra', *South European Society and Politics*, 9 (2): 132-158.

Moura, D. *et al.* 1978. *O que é a energia nuclear: oportunidade em Portugal*, Lisbon: Moraes Editores.

Oliveira, J. C. 1977. *A energia nuclear: bases para uma opção*, Lisboa: Sá da Costa, 1977

—. 2002. *A energia nuclear em Portugal: uma esquina da história*, Santarém: O Mirante.

Oliveira, J. M. C. & E. J. Martinho. 1978. *Energia nuclear: origem e aplicações*, Lisbon: Junta de Energia Nuclear.

Paiva, J. 2008. "Marcos ambientais da década de 70", In *60 anos pela natureza em Portugal*, Lisbon: Liga para a Protecção da Natureza.

Pampel, F. C. 2011. "Support for Nuclear Energy in the Context of Climate Change: Evidence From the European Union." *Organisation & Environment* 24 (3) (October 4): 249–268.

Pereira, T. S., A. F. Rodrigues, A. Carvalho & J. A. Nunes. Forthcoming. "Parlamento e Conhecimento Científico: Dupla delegação?", In *Democracia, Deliberação e Participação Pública,* ed. J. M. L. Viegas, Lisbon: Mundos Sociais

Pinto, P. R. 2008. "Urban Social Movements and the Transition to Democracy In Portugal, 1974–1976". *The Historical Journal,* 51, 1025-1046.

Ribeiro, A., F. Barriga & J. Cabral. 2008. "Sismotecnónica e Segurança Nuclear: o caso do Douro Internacional". *Geonovas*, 21: 11-14.

Rodrigues, J. N. & V. Azevedo. 2006. *Nuclear: o debate sobre o novo modelo energético em Portugal,* Famalicão: Centro Atlântico.

Rucht, D. 1997. "The impact of anti-nuclear power movements in international comparison". In *Resistance to new technology*, ed. M. Bauer, Cambridge, Cambridge University Press: 277-292.

Ruivo, B. 1998. *As políticas de ciência e tecnologia e o sistema de investigação*. Lisboa: Imprensa Nacional Casa da Moeda.

Santos, B. S. & J. A. Nunes. 2004. "Introduction: Democracy, Participation and Grassroots Movements in Contemporary Portugal", *South European Society and Politics*, 9 (2): 1-15.

Schmidt, L. 2008. Ambiente e Políticas Ambientais: escalas e desajustes. In *Itinerários: A Investigação nos 25 Anos do ICS*, eds. M. V. Cabral, K. Wall, S. Aboim & F. C. Silva. Lisbon: Imprensa de Ciências Sociais: 285-314.

Schofer, E. 2003. "International Science Associations, 1870-1990". In *Science in the Modern World Polity: institutionalization and globalization*. Eds. G. S. Drori, J. W. Meyer, F. O. Ramirez, & E. Schofer, Stanford CA: Stanford University Press: 81-99.

Scott, J. M., J. L. Rachlow & R. T. Lackey. 2008. "The Science-Policy Interface: What is an Appropriate Role for Professional Societies?" *BioScience*, *58*(9): 865-869.

Soromenho-Marques, V. 2003. "O Associativismo ambiental em Portugal: Potencialidades e limites", in *Novas Formas de Mobilização Popular*, ed. J. Rebelo, Porto: Campo das Letras: 197-207.

Teixeira, L. H. (2011) *Verdes anos : história do ecologismo em Portugal (1947-2011)*, Lisbon: Esfera do Caos.

Torres, C. M. S., & W. Rudig. 1983. "Nuclear Power in Argentina and Brazil." *Review of Radical Political Economics* 15 (3) (September 1): 67–82.

Touraine, A. Z. Hegedus, F. Dubet & M. Wieviorka. 1980. *La prophétie anti-nucleaire*, Paris: Seuil.

Weart, S. R. (1988), *Nuclear fear: a history of images*, Harvard: Harvard University Press.

Wynne, B. (2002), "Risk and environment as legitimatory discourses of technology: reflexivity inside out?", *Current Sociology*, 50 (3): 459-477.

CONTRIBUTORS

Sandra Assis is an Anthropologist, currently finishing her PhD in Biological Anthropology at the Department of Life Sciences of the University of Coimbra (Portugal). She is a member of the CIAS Research Centre for Anthropology and Health from the University of Coimbra, and a founding member of GEEvH Group of Studies in Human Evolution. She has published several scientific articles in national and international journals and a science book for children that explores the importance of the human skeleton, as well as the work of anthropologists and archaeologists. Her research interests include the application of histological techniques in the study of periosteal new bone formation in human skeletal remains. She has also participated in the development and promotion of recreational and pedagogical activities for children in the field of Biological Anthropology.

Cristiana Bastos (CUNY 1996) is an anthropologist with an interest in the history of science, technology and medicine and in transnationalism. She is a senior researcher at the Institute of Social Sciences, University of Lisbon, and has taught in different graduate programmes in Portugal, Brazil and the United States. Her publications include *Global Responses to AIDS: Science in Emergency* (Indiana University Press, 1999), *Clínica Arte e Sociedade* (Imprensa de Ciências Sociais, 2011), *A Circulação do Conhecimento: Medicina, Império, Sociedade* (Imprensa de Ciências Sociais) and assorted articles in *Identities, International Migration, Bulletin of the History of Medicine; Journal of Southern African Studies, Anthropology and Medicine, Manguinhos História, Ciencias, Saúde, Etnográfica, Análise Social.* She is currently preparing a monograph on colonial medicine in Goa and another on the racialisation of migrants.

Vanessa Campanacho is an Anthropologist, currently doing a PhD in the Department of Archaeology of the University of Sheffield (United Kingdom). She belongs to CIAS (Research Centre for Anthropology and Health) of the University of Coimbra, Portugal and to GEEvH Group of Studies in Human Evolution. Since 2008, she has authored and co-authored several articles, such as the article entitled "Assessing the Influence of Occupational and Physical Activity on the Rate of Degenerative Change of the Pubic Symphysis in Portuguese Males from the nineteenth and twentieth Century", published at 2012, in the *American Journal of Physical Anthropology.*

Cristina Palma Conceição is a Portuguese sociologist. She has a PhD in Sociology and an MSc in Communication, Culture and Information Technology, both from ISCTE-IUL (University Institute of Lisbon), where she currently holds a position as assistant professor. Since 1998 she has been involved in several research projects, mainly in CIES-IUL (Centre for Research and Studies in Sociology). As a researcher, she specialises in the social studies of science and technology. Her main areas of interest are scientific culture and public communication of science, science as a social institution, and technological innovation.

Ana Delicado is a Research Fellow at the Institute of Social Sciences of the University of Lisbon. She holds a PhD in sociology (2006). She works mainly in the field of social studies of science. She has carried out research on non-governmental organisations and volunteering, environmental risks, museums of science and public understanding of science, the use of the internet by children, international scientific mobility, the climate change research community and scientific societies. She is currently coordinating a research project on renewable energies. She is vice-coordinator of OBSERVA Observatory of Environment and Society. She has published four books, 16 book chapters and 22 articles in peer-reviewed journals.

Dulce Freire is a Research Fellow at the Institute of Social Sciences of the University of Lisbon. She has a degree in History (1992), a Master's degree in History (1998, whose dissertation was awarded the *Mário Soares Foundation History Prize*) and a PhD in Economic and Social Contemporary History (2008), all from the Faculty of Social Sciences and Humanities, New University of Lisbon. Since the beginning of the 1990s, she has been conducting research in subjects related to rural and agricultural history in Portuguese and Iberian contexts. She has been analysing questions such as public policies and institutional frameworks, social movements, memory, technological innovation and social change. Seeking to develop interdisciplinary research, she has been a member of several teams and she has been dealing with different theoretical and methodological tools that could be useful for establishing links between written, oral and visual evidence. She was a research fellow at the Institute of European Studies, University of California at Berkeley (2001), and she did other research visits at the École de Hautes Études en Science Sociales (2005) and at the Universidade de Santiago de Compostela (2006).

Luís Junqueira is a PhD candidate in Sociology at the Institute of Social Sciences, University of Lisbon. He has a Master's degree in Sociology from the Faculty of Social Sciences and Humanities, New University of Lisbon. Over the last few years he has worked as a research assistant in projects in the area of the social studies of science.

Rui Luís is a speleologist and member of AESDA. He is the editor of *Trogle* and has authored several scientific papers.

Ana Cristina Martins is Research Fellow at the Institute of Tropical Scientific Research, under the Science 2007 programme, where she also engages in projects in the area of the History of Science in general, and the History of Archaeology in particular. She has a PhD in History, a Master's Degree in Art, Heritage and Restoration and a Bachelor's Degree in History–Archaeology from the University of Lisbon, where at the institution's Archaeology Centre–Uniarq–she developed a post-doctoral project on Archaeology in Portugal between the 1920s and 1960s, and was principal investigator of the line "History of Archaeology in Portugal. Theoretical Issues". She has published several works on the History of the evolution of thinking regarding archaeology, museums and heritage, most of which result from papers presented at national and international meetings. She teaches at the Lusophone University of Humanities and Technologies (Lisbon), as Invited Lecturer, coordinating the History of Heritage and Science Section that forms part of the CPES-Centre for Research and Social Studies of the Faculty of Social Sciences and Humanities. She also chairs the Archaeology and Heritage Studies Sections of the Geographical Society of Lisbon.

Rui Mergulho is a speleologist and member of AESDA. He has authored several scientific papers.

Ricardo Moreira holds a BA in Anthropology from the University of Coimbra, having written his first ethnographic work on rural history in the South of Portugal. In 2010, he joined the Institute of Social Sciences, both as a graduate student and as a research assistant, taking part in the *Nuclear Portugal* research project. Since then, he has engaged with the history of Nuclear Physics in Portugal and its role in the emergence of cancer research in Portugal. He is presently writing an MA dissertation in Social and Cultural Anthropology on this topic.

Paulo Pacheco is a speleologist and member of AESDA.

Hugo Pereira was born in Oporto in 1979. In 2005, he completed his History undergraduate programme in the Arts and Humanities College of Oporto University. Three years later he completed his Master's programme in Contemporary History at the same institution with an investigation about the relationships between the lower house of the Portuguese parliament and the construction of railways between 1845 and 1860. He has a PhD degree since 2012 with a dissertation about Portugal's railway policy in the second half of the nineteenth century. He published several papers about the History of Portuguese railways

Rui Pinto was born in 1962. He is a teacher and works also in teacher training. He has a degree in History of Art (1986) by the University of Lisbon and a Master's degree in History of the Discoveries and the Portuguese Expansion (1994) by the same university. He finished his PhD thesis, entitled "Gago Coutinho (1869-1959), geographer and historian: a scientific biography", at the same university. He has published books and dozens of articles in Portuguese and foreign journals. He is a member of the Naval Academy, President of the Section of History of the Lisbon Geographical Society, a researcher at the Centre of History of the University of Lisbon and at the Joaquim Verissimo Serrão Research Centre, a corresponding member of the Historical and Geographical Institute of São Paulo and of Espirito Santo, Vice-President of the Lisbon Study Committee of the Portuguese Archaeologists Association. He has organized several scientific meetings.

Frederico Tátá Regala is a speleologist and President of AESDA. He has a Master's degree in Archaeology by the University of Algarve. He has authored several scientific papers.

Raquel Rego is a sociologist and a Research Fellow at SOCIUS-ISEG, University of Lisbon. She has been carrying out research about associations (consumers, professionals, workers, employers). She is the author of *Dirigeants Associatifs: Engagement et Professionnalisation*, published in 2010 by Harmattan, and the editor of *The Trend Towards the European Deregulation of Professions and Its Impact on Portugal under Crisis*, published in 2013 by Palgrave Macmillan.

INDEX